Daniele Gasparri

Oh Wow!
I saw Saturn!

Guide to observe the night sky for young people and adults

Copyright © 2018 Daniele Gasparri

ISBN: 9781731566829

This work is protected by the Copyright Law. All rights, in particular those pertaining to reprinting, translation, the use of images and tables, oral quotation, radio or television broadcasts, reproduction on microfilm or in databases, different reproductions in any other form, paper or electronic, are reserved even in the case of partial use. The reproduction of this work, or any part of it, is allowed within the limits established by the Copyright Law.

Translation: Mary Purpari

Cover image: A composition of two photos taken by the author. The center of the Milky Way seen from the incredibly dark sky of Atacama desert, northern Chile, and the planet Saturn, imaged through a 9.25 inches telescope. These are among the most beautiful wonders of the sky, but, of course, they are not the only ones.

Table of Contents

Introduction 1
Greek Alphabet 2
Falling in love with the heavens 3
    Why use a telescope? 8
    Don't be fooled 10
    Under a perfect sky 12
    How difficult is it? 21
The Sky seen with the naked eye 23
    What can be observed with the naked eye? 25
    The light pollution 26
    What are the constellations? 29
    A special star: the North Star 31
    The Sky that changes 33
    Let's get our bearings in the sky 36
        Let's find the North Star 37
        The magnitude 40
    The first date with the stars 41
    Let's recognize the easiest constellations 43
    The sky in spring 45
    The summer sky 49
    The autumn sky 52
    The winter Sky 55
    The first observation 59
    Not just stars 74
        Falling Stars 74
        Satellites 75
    The zodiac constellations 77
    Finding the Planets 79
    Searching for the planets 85
    The rare surprises, the eclipses 86
    A help in the field: amateur astronomers and star parties 91
The sky through a binoculars 93
    How are binoculars made and how do they work? 93
    The first attempt with your binoculars 95
    Preparing the first "official" evening with the binoculars 95
    How to observe with binoculars 97
    How do we find objects on the map? 97
    A few other useful techniques 99

Where is north and south? 101
Our first "official" observation 102

The Telescope **115**
What is a telescope? 116
Why does a telescope let us see better? 116
The various telescopes 118
The mounts 120
   The types of mounts 121
A few numbers on telescopes 124
Choosing the telescope 125
   Which telescope is right for us? 126
   What about automatic pointing? 128
   Which brands should we choose? 129
   From whom should we buy the telescope? 130
How do we make the telescope work? 131
   What do we need for observing? 132
   Focusing 133
   Aligning the finder 134
   Upside-down images 135
   The movements of the equatorial mount 136
   Aligning the equatorial mount to the pole 137
   Balancing the telescope 139
   The Acclimatization 140
   How to set up the GOTO? 140

The Big Moment: let's observe with the telescope **141**
Some advice for better observations 143
Observing the Solar System 146
   What can we observe on the planets and how? 151
   The Sun and the Moon 159
   Asteroids and Comets 166
   Advices for better observations of the planets 167
   Why observe the planets? 170
The deep sky 172
   What's the point of observing nebulae, galaxies and star clusters? 172
   Where are deep-sky objects found? 173
   To each season its objects 175
   Double Stars 176
   Open clusters 178
   Globular clusters 184
   Nebulae 190
   Galaxies 198
   Observing the arms of spiral galaxies 200

 Some particular observations 206
  The colors of the stars 206
  Alien planetary systems 207
  The farthest object 209
  The Messier marathon 211
 Keeping a record of our sky 212
 When art meets astronomy 213
 Unfettered curiosity 219
Appendix **221**
 The Collimation 221
 Cleaning the optics 222
 Most frequently asked questions: a quick review 223
Biography **237**

Introduction

When I begged my parents to give me a telescope for Christmas way back in 1993, I didn't even know where to look for the stars.

My childish unawareness made me take a much greater step than I could have ever imagined; but, at the same time, I pushed myself enough to dare to undertake a path fraught with obstacles.

Now, almost twenty years after I first set up my small, and very beautiful, refractor in my grandparents' kitchen, things have changed. I can finally say to myself that I have completed the very long path that I began as an ingenuous dreamer and have finished as an amateur astronomer. I now know how to use a telescope; now I know how it works; I know how to choose the right one for my needs; I know how and what to look for and, above all, what to expect. It took me almost twenty years to learn what little I know and this is the reason that I wrote a book: to help other enthusiastic young people to avoid waiting so long to knowledgably, and satisfactorily, observe the wonders of the Universe. Because you don't just suddenly become amateur astronomers, but you become such after a time during which you must necessarily learn how to move in the immense Universe above our heads.

Me and my first telescope. Christmas 1993; I was 10 -years old.

Making mistakes is the best way to learn, so I don't want to deprive anyone of this great opportunity. However, I would like to give some useful advice so that your mistakes take you in the right direction, without having to wait twenty long years.

Partly autobiographical and often ironic, we'll go on to the discovery of amateur astronomy and the magnificent Universe that is right above our heads. It begins at just 60 miles above us. For now, we can only hope to get close enough to understand some of its secrets and, most of all, get excited about it, through the use of a telescope.

Because, there are only two words comprising the amateur astronomer's password: getting excited, without any limits.

Daniele Gasparri
November 2018

Greek Alphabet

name	letter	spelling	capital
Alpha	α	ἄλφα	Α
Beta	β	βῆτα	Β
Gamma	γ	γάμμα	Γ
Delta	δ	δέλτα	Δ
Epsilon	ε	ἒ ψιλόν	Ε
Zeta	ζ	ζῆτα	Ζ
Eta	η	ἦτα	Η
Theta	θ	θῆτα	Θ
Iota	ι	ἰῶτα	Ι
Kappa	κ	κάππα	Κ
Lambda	λ	λάμβδα	Λ
Mu	μ	μῦ	Μ
Nu	ν	νῦ	Ν
Xi	ξ	ξῖ	Ξ
Omicron	ο	ὂ μικρόν	Ο
Pi	π	πῖ	Π
Rho	ρ	ῥῶ	Ρ
Sigma	σ ς	σίγμα	Σ
Tau	τ	ταῦ	Τ
Upsilon	υ	ὖ ψιλόν	Υ
Phi	φ	φῖ	Φ
Chi	χ	χῖ	Χ
Psi	ψ	ψῖ	Ψ
Omega	ω	ὦ μέγα	Ω

Falling in love with the heavens

Me and billions of stars. A 30 seconds exposure taken from the darkest sky of the world: the Atacama desert, Chile. I choose to live here; now you know why.

The Universe was defined – it's not known who defined it first – as everything that exists. It's a beautiful definition, but it doesn't tell us much and it certainly doesn't give us the push that is necessary for wanting to explore it.

So then, what is the Universe in reality? And why should we explore it with our telescope?

It's highly likely that if we are reading these words in a book about astronomy, we are already curious and passionate enough to discover it. But are we really aware of the wonders we can observe at night? In final analysis, if we lift our eyes toward the sky from any one of our cities, all that we'll see is a black expanse illuminated here and there by some tiny dots called stars.

So, at a first glance, this sliver of the Universe we can admire (so to speak!) every evening with the naked eye seems anything but interesting and exciting.

What do we hope to find in a cesspool?

Actually, what we are observing isn't the sky, but one of the greatest havocs wreaked by modern man in the name of an arrogant technological growth. If the sky seems to be without interest, we owe it to the myriad of artificial lights always turned on that have literally turned off the thousands of stars that it would be possible to admire.

No, I didn't make a mistake and I'll repeat it: thousands of stars.

The best gift we can give ourselves is choosing a Moonless night, go far away from the city and big towns' lights, even better in the hills or mountains, and lift our eyes to the sky, the real one. If we've never tried it before, it will be an indescribable sensation. That dark and uninteresting place is now, right here in front of us, illuminated by thousands of tiny stars: it's unbelievable!

On a dark night, without those tiresome lights all around us, we can observe at least 3,000 and even up to 5,000 stars filling every space and creating the impression of a crystalline dome, in which we'd like to navigate for the rest of our life.

During summer nights, the best ones for observing thanks to the mild temperatures of the nights, it's possible to observe something that we would never have thought possible. In fact, the stars aren't put randomly in the sky, but seem to be concentrated in a zone that starts almost above our head and ends to the south, until it touches the horizon. We're observing the Milky Way, our galaxy, an immense island of stars that finally shows up in all of its glory: 200-400 billion stars arranged in (almost) perfect order.

Let's sit on the ground, in silence, and contemplate the sky. Maybe we'll see some little dots moving from time to time and UFOs will immediately come to mind; other times we'll see what are called falling stars, quickly speed through the sky. We'll surely have more questions than answers at this point, but there's time to learn.

All we have to do is to think about this window to the Universe, and understand how beautiful and profound it is. Yes, because those miniscule dots called stars, which the ancient civilizations thought were lights hung from a large vault by the gods, are actually absolutely fascinating objects. They are immense gaseous spheres, 100, 1000 and even 10,000 times larger than our planet, that shine with their own light. A small portion of this light has travelled an incredible distance and is resting, right this moment, on our eyes so they can be filled with these little cosmic lamps that fill the entire sky.

Stars, many, very many, many of which are similar to our Sun. Yes, the Sun is a star and not even one of the biggest; on the contrary, it's fairly small. At 92.96 million miles from us (mama mia!), it is the source of life with its heat.

92.96 million miles are an incredible distance to us, and yet, it's a very short distance for the Universe. If the light of the night-time stars seems weak to us compared to the Sun, how far away are they? The answer is simple: farther away than we can possibly imagine.

The summer Milky Way, captured by a digital camera, in a dark sky that show it in all of its glory. Even though we can't see any of these contrasts or their colors with the naked eye, it is, in any case, an impressive view of the power of this extraordinary Universe.

So then, we speak of light years, but few really know what this strange unit of measure means. A light year is the distance covered by any ray of light in the vacuum of space, equal to 5.8 trillion miles.

The nearest star is around 4.3 light years away, but this small, familiar number entails a distance of almost 25 trillion miles!

Almost all the stars that we can admire at night are more than 10 light years away; some are 100 and others still, more than 1000. And if the vastness of the cosmos isn't enough to leave us speechless, light years tell us something else that's spectacular. Light is the fastest moving thing in the Universe. As hard as we might try, Nature's own rules tell us that nothing can go faster.

Of all the objects in the heavens, we see the light, whether it's reflected (like with the planets) or emitted (stars, nebulae and galaxies). But if its speed, even though very high, is finished, whether it takes years, centuries or millennia to reach us, it means that the Universe is a gigantic time machine. If a star 10 light years away sends us light, it will travel through space for that amount of years before reaching the Earth, our eyes and telescopes. When we receive the information of this star's existence, it is by now ten years old: we are, therefore, seeing the star as it was ten years earlier.

Light is what travels the fastest in the Universe, so we'll never be able to observe an object exactly as it is when we are looking.

The Andromeda galaxy, the closest to us, is about 2.3 million light years away, so we are observing it as it was the same amount of years ago. Perhaps many of its blue stars (the ones who live the least) no longer exist according to Andromeda time, while they are still very much alive on our timeline.

Reversing the story, if a technologically advanced species living on some planet orbiting around a star in the Andromeda galaxy had developed a telescope powerful enough to observe the Earth, he would have no idea that this planet is inhabited by intelligent creatures. From their position, the events of the Earth would be 2.3 million years behind. So, at the most, they would see the first hominids hunting on the African steppes.

Going further into space, an observer from 5 billion light years away wouldn't even see the Sun and the Solar System, which would be born only half a billion years later.

The further we look into space, the further back we look in time. There's no sense in saying that what we actually see is an image that doesn't correspond to reality. Contemporaneousness is a concept that doesn't exist in the Universe, or at least according to our common definitions.

Isn't this fascinating? We have the entire history of the Universe in front of us, clearly visible with our telescopes and certainly much more fascinating than what we read in some boring schoolbook. And if all this still isn't enough, here is the most profound and intimate answer of why we should always look up: because we come from there, because our history and the answer to all of our questions is also written there, just 62 miles above our heads.

The Earth, as big as it is, is only a drop in an immense ocean that is just waiting for us to explore it. We, with our eyes, our telescope and our desire to discover, are the necessary ingredients for taking off on the most beautiful, fascinating and surprising voyage of our entire existence. We are inhabitants of the Earth but citizens of the Universe.

Do we still need to look for reasons for loving the Universe and wanting to explore it with our efforts? We have an enormous window that opens up onto the most beautiful panorama that exists; do we really need to look for reasons to look through it? Instead, what we'll really need is to find the strength to pull away from it, because it could be that we'll never want to look away again!

The Andromeda galaxy is the farthest object visible to the naked eye. The light that we see in the sky and in this photo is 2.3 million years old. Mankind wasn't even on the Earth yet when it left. And so, the Universe is an immense time machine: the further we look out into space, the further back we go in time. And naturally, the other way around is also true: a planet in Andromeda would see the Earth as it was 2.3 million years ago, and there would be absolutely no trace of our existence.

Why use a telescope?

The answer is simple: because, although the evening sky already excites us as seen with the naked eye, we'll be able to discover objects and secrets that we'd never be able to observe with our eyes alone.

Because those faint stars that during the summer nights seem to prefer one corner of the sky rather than another, aren't the only objects in the Universe; neither are the Moon, the Sun and a few falling stars here and there. The Universe, the real one, is explored with a telescope. Not NASA's really powerful ones, but our small instrument – sometimes not much more than a toy – is sufficient to show us something that we would never have thought of discovering.

Maybe a telescope wouldn't even be necessary to increase our desire to explore. Of course, a little curiosity and a binoculars might be enough, just like happened to me a long time ago.

I had just turned ten and with the insistence of a child that age, I forced my father to buy me a pair of binoculars like the ones I admired in the American movies. I didn't know what to do with them, but the idea of being able to magnify things, just like I had seen on TV, fascinated me. However, I still didn't know the sky at all, so I always used them during the day to admire the panorama.

Then, one night at the end of the summer, the crescent Moon that shone into the living room through the window, while my grandmother was cooking dinner in the kitchen, called to me for no apparent reason. I had my binoculars with me and I said to myself, "If I can magnify the panorama, what would happen if I aim it at the Moon? Will it look bigger?" A fraction of a second between thinking and the reckless action of a curious child, and my life changed forever.

That evening, I sat there for hours looking at the Moon, enchanted, my mouth wide open, admiring that surface full of "holes" that I would never have expected to find, in total silence. I can still feel those emotions as I write these lines, seeing myself and feeling so small, with my heart in my throat as I incredulously rub my eyes, trying to understand if what I'm seeing is real or not. Running to my grandmother to let her see that wonder, insisting that my parents tell me what those holes were and the desire to dream, to fly, to explore, to discover, that I felt in my young soul for the very first time.

It was the beginning of a love that would never die, of a relationship, at times conflicted but always full of respect, between me and the Universe, between me and our origins, between me and the reality surrounding us – which certainly isn't made of either Iphones, let alone stupid videogames whose only purpose is to make us stop dreaming.

And so, why use a telescope? Because, if the Moon's craters are only the most evident things, visible even with just binoculars, we need an apparatus much more powerful than the eye for the real treasures. If we are excited just by observing those tiny dots called stars, discovering what is hidden even deeper will leave you speechless. And there's no immediate need to go far into space and time to be excited.

The planets in the Solar System are spectacular. Jupiter shows imposing cyclones in rapid evolution and four of the numerous moons that orbit around it. Mars can show faint clouds, superficial spots and one of the two polar icecaps composed mainly by water-ice, even with small telescopes. Let's stop for a moment: we're admiring another word, more than 30 million miles away, on which primitive life forms (bacteria) could even exist. And what can we say about the planet par excellence, the one that very often keeps even the most experienced observers glued to the telescope for hours? Naturally, I'm talking about Saturn, an incredible globe surrounded by a spectacular ring. There will be time later to talk about my feelings when I observed Saturn for the first time. Now, I can only say that during evenings with the public, even the least interested suddenly change their attitude when they observe Saturn through my telescope. After a few seconds, during which they can't express words, the first thing they say 90% of the time, while looking at me with shining eyes, is this: "Tell the truth; you put a photograph in front of the telescope. It can't be that beautiful!"

Well, yes, it is beautiful, very beautiful and, together with Saturn, so is the entire Universe. Because, going beyond our Solar System we can discover star clusters, nebulae and even galaxies.

To see these objects, and I'll repeat it more than once, we need a very dark sky, without even a crescent Moon. If conditions permit, we'll push forward to thousands or millions of light years from the Earth, an unimaginable and absolutely fascinating distance. We'll see more concentrated balls of light, often very faint, in the center. And, if by chance they tell us anything from an esthetic point of view, let's think that that is the light of a galaxy, a gigantic island where there are hundreds of billions of stars, billions of planets, thousands of nebulae and star clusters. We are observing a real universe in the Universe, much larger than what our imagination could ever conceive. After almost twenty years that I have been working with all this, I can still get excited by just repeating the same things for the thousandth time. Because you can never get tired of the Universe.

Don't be fooled

Observing the sky with your own telescope, with your own efforts; going around without limits of time and space; feeling that you are in direct contact with the Creation and experiencing the greatest freedom of all times. These are the wonderful emotions we feel from observing the Universe through a telescope, our telescope. Emotions that no photograph observed on the cold screen of a computer could ever give us.

With the widespread use of internet, it's now very easy to surf the web searching for spectacular astronomical images. Some leave us breathless as to colors and details. Well, let's clear things up immediately: this isn't astronomy and we'll never, ever have these views with a telescope.

The direct observation of the sky reaches emotions that are different compared to the impact of an astronomical photograph. If we hope to see through the eyepiece of our telescope as well as an image on our computer's screen, we'll unfortunately be sadly disappointed. I'll repeat it once again, just to be clear: no telescope will show objects as they appear in a photograph. It's not the instrument that is limited, so it's useless to go out and try to buy the Hubble Space Telescope: even that one, for example, will show almost all objects – except the brightest planets and stars – without color.

The limits are with our eye which, although extraordinary, certainly hasn't evolved enough to let us admire the beauty of the Universe. During the millennia of difficult natural selection, our apparatuses, including visual, were developed to simply guarantee us with the maximum probabilities of survival in an extremely hostile world.

One of the biggest limits of our eyes in astronomic observations is represented exactly by the extreme difficulty in seeing the colors of the astronomic objects. And in these cases there is no powerful telescope that holds: almost all star clusters, all galaxies and a good part of the nebulae, will always appear in black and white.

If "low power" instruments are used (further ahead we'll see what that means and how a telescope's power is measured) another problem is the lack of contrast, above all with nebulae and galaxies. This translates into objects that become similar to tenuous little clouds instead of the splendid curtains of gas admired in the photos.

The planets will also seem smaller, with fewer particulars, but in this case a good part of the blame will be attributable to our still meagre experience. We'll have time to talk about this, too. For now, let's end by underlining the concept: an observation made through a telescope is completely different respect to what it can be seen in a photograph. It's different because the details observed are completely different (to the photograph's advantage); it's different, though, especially because of the emotions, because there's no photograph that can stand up to the unique sensations that we feel when we observe the same object with the telescope, with our own eyes. No digital camera is able to create the unique tie with the Universe than we could generate by observing with our limited, but extraordinary, eyes.

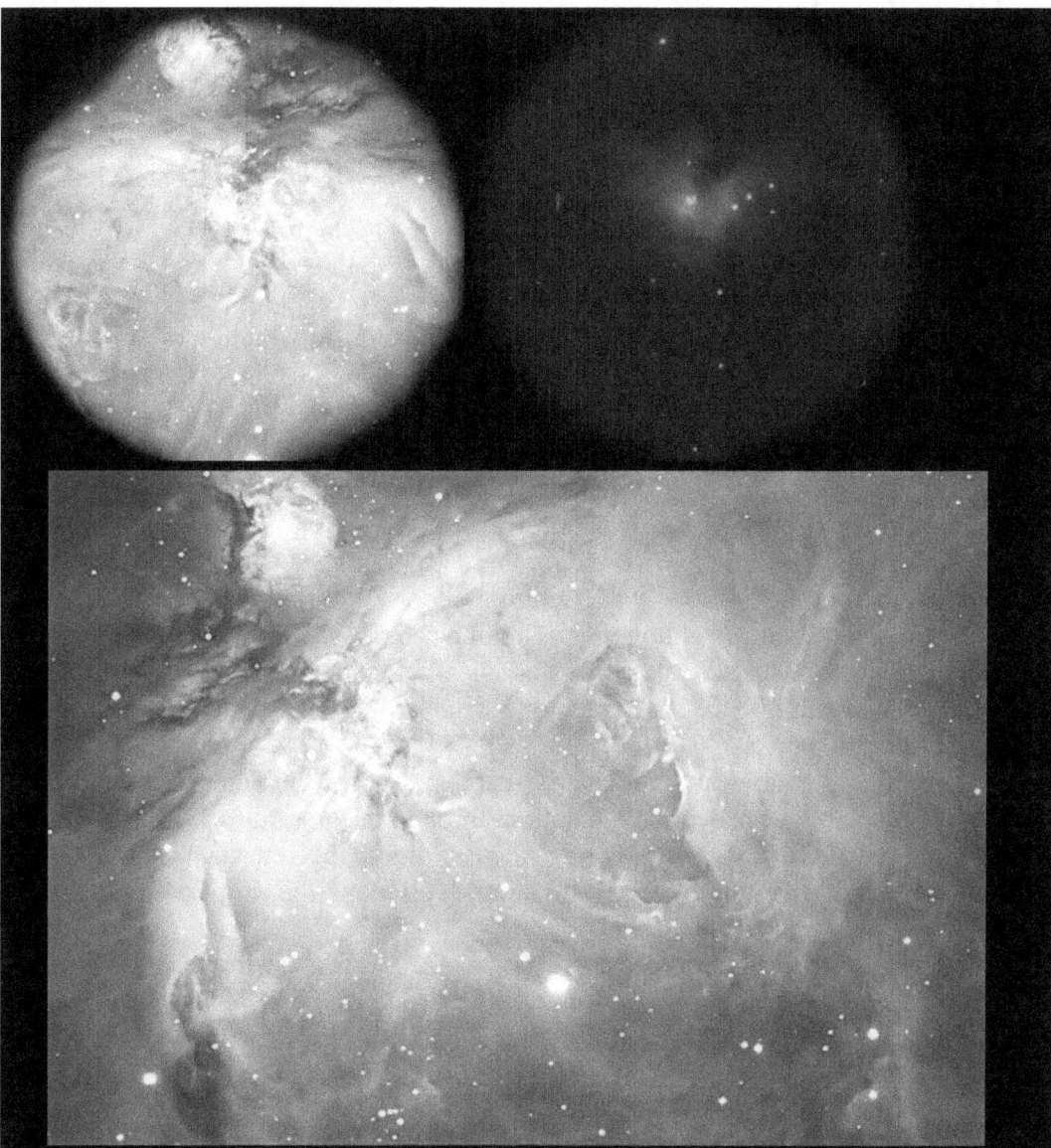

Unforgiving comparison between the production of a digital camera and the human eye with the same instrument and sky conditions (with medium pollution) on the great Orion nebula. **Above:** focus on the same area between the view through a digital camera (at left) and the eyepiece (at right). **Below**: the real extension of the Orion nebula is shown, which is impossible to observe through the eyepiece of any instrument. If the only reason we want to observe the sky is to see the beautiful images given on Google, then it's better to stick with Internet. Amateur astronomy through a telescope is fun for other reasons besides the mere display of a photo. It's the contact with the Universe, it's the thrill of exploring an immense world, it's the crowning of the dream to reach the stars and understand our origins. It's pure knowledge, immersion in an extraordinary Universe. It's passion for life and for the Universe that goes far beyond a brushstroke of incomprehensible colors. And if we don't use the images from the Hubble space telescope as a referral, many objects are spectacular with our small telescopes. The important thing is to not have mistaken expectations.

Under a perfect sky

Before going into the more practical part, I'd like to jump the gun a little and describe an experience that I had, after observing the skies for almost twenty years. It was an experience that I never thought would be so amazing and testifies to how one can never be tired of seeing the Universe.

From October 31 to November 20, 2012 a dream that I'd had ever since I was a child was fulfilled: visiting Australia, travelling between the big cities and the endless clearings, until I found the desert and observed the Sky, with a capital S; that wonder that only our grandparents can remember here in Europe, cancelled now by the wild and completely uncontrolled havoc wreaked by public illumination.

The Sky – that environment which from our cities often appears orange and lacking of interest because it is populated, at most, by a handful of stars – lights up as the greatest and most exciting display that we could ever hope to see, in a dark place, hundreds of miles away from the nearest big cities.

The stars come out into the open. At first there are hundreds and then several thousand… Together they are able to weakly illuminate the environment around us, but it seems dark, so dark that it's almost impossible to see our feet or the hand we hold in front of us. Shining with a weak glow, this uncontaminated sky unveils at our eyes gems and phenomena that we have often only read about, with a good dose of skepticism, in astronomy books.

The zodiacal light, that weak glow that is visible only after the Sun has set, or before sunrise, produced by the reflection of sunlight through dust particles present along the ecliptic plain in the Solar System, is so evident it almost becomes annoying. A long cloak of white light rises up to the zenith, at times all along the ecliptic, from horizon to horizon.

"Man, those are the lights from a big city!" is the expression that each one of us, including me, would exclaim with dismay the first time we observe it.

Then, our rationality returns. "No, that's not possible; I'm a hundred twenty-five miles from the nearest town, 1,875 miles from the big city; that can't be an artificial light!"

And so, the perfect night begins, amazing us with the natural lights of the sky. And then, if the clarity of the zodiacal light sets over the center of the Milky Way, about 15° high over the horizon, the view is thrilling: something that no book is able to describe and no one can imagine it until they see it and feel it with their own eyes.

Each of the 5 nights spent in the Australian outback always begins like this, in the company of curious kangaroos hopping around, warily drawing closer to those strange two-legged animals that make stupid sounds while observing what is for them the most common of things existing.

Every night begins like this, both from the point of view about the display in the sky, as well as the emotions, too great and hidden for too long to be able to calm down in just a few hours, compared to the years spent under a stupid cloak of artificial light.

After the initial amazement, we recover enough to begin looking at the sky and we immediately discover that here, the constellations visible also in our sky are… upside down! It's normal: here, we are literally upside down with regards to our latitudes, our rational part suggests. And yet it's a sensation that is so strange; I would say funny, if it weren't for the discomfort one feels trying to recognize, often in vain, upside down figures!

There's no time to be annoyed; in fact, one smiles because he knows perfectly well that this is only an interlude between two big emotions: the first, already over, is from the zodiacal light, while the second is yet to come and is represented by the treasures that are visible only in the southern hemisphere, those brilliant and surprising gems that our northern horizon will hide from us forever.

Our gaze then runs toward the south, because that's where the invisible sky is.

A quick look and right off there's a disappointment: "There are two clouds in the middle; look how sparkly and annoying they are. How unlucky!"

Light pollution? Maybe not…

At first, we wait for a few minutes, hoping that they'll go away, but it doesn't happen: "You're tough, you haven't moved an inch!"

You could go on waiting for an eternity, if it weren't for one extremely important detail: the stars in the vicinity of the horizon are disappearing and a disturbing dark shadow is eating the weak glow of the sky.

At this point, an unconscious doubt begins to worm its way into my mind, struggling against denial to reach conscious thought. The words I read once in an old astronomy book come to mind: "Clouds seem darker in an uncontaminated sky and, at times, appear to be real black holes."

"Absurd!" I exclaim to no one in the dark, silent night… "I can't believe that the pitch black ones are really clouds! I've never seen them like that!"

"But then, what are those shiny clouds above them, so similar to the illuminated ones in our skies?" If those burning, disturbing eyes that watched me there in the middle of the meadow had been aware, they would probably have been amused as they exclaimed, "Poor fool, do you live under a rock? They are clouds, but of the Universe; they're called Magellanic clouds!"

From that moment on, my eyes, put to the test by so many emotions all at once, will open up wide as though to gather in as much light as possible, and won't stop looking all night, every night, from that incredible work of art above our heads, which we call the Universe…

And yet, we don't have to admire a portion of the sky that we'll never see here to be amazed by the Universe. Let's try to understand what we would see in our sky if, someday, we were to go to the middle of Spain or United States (for example the Bryce Canyon or Wyoming), where the sky is still dark enough to see the real clouds darker than the stars.

Clouds in the Universe: the Magellanic Clouds.

What follows next is the diary of an evening spent with a few friends in a wilderness area of Australia, far from any source of artificial lighting. It was the first time I saw a perfect dark sky. Now that I live in the Atacama desert, under the darkest night sky or the whole Planet, I can see this beauty whenever I want, at least 320 clear days every year.

...My dinner was interrupted after the first quarter of chicken by a guest that we had waited for at length, and which I'll never get used to: the light of the zodiac.

I see it perfectly from the hood of my car, despite the headlamps on Marco's and Francesco's forehead.

I try to wash my hands with a napkin, and then I delight Marco: "Look, you can see the zodiac light perfectly; I'm going to photograph it. Are you coming, too?"

Completely awed, they turn toward the southern horizon and, convinced that they're observing the normal light of a big city, they ask me, through Francesco: "Help me out here; I only see a strong light, so where is the zodiac light?"

How quickly the perspective changes: just a few days ago, this was the phrase that crushed me from within, disoriented and baffled at how evident that column of light was... Now, I had become the guide.

As usual, I had to learn that little that I know, by myself, among a thousand efforts but also with enormous satisfaction. And now I'm the one giving out advice about the Southern sky, which makes me as proud as a son is of his father.

"That light there is exactly the zodiac light! It's not the sunset because it's white; it's not even the light pollution produced by some city, because there aren't any around here. This is natural light pollution!"

"Fantastic, unthinkable!", "Unbelievable!" "It really does look like the city lights; it's jaw-dropping!" exclaim Marco, Francesco and Malù almost at the same time.

I can't hear anything else because Marco is literally dragging me up the hill. "Let's go take pictures, Daniele! What are we waiting for?"

We quickly leave the "table" and let the light in the sky guide us toward our equipment; their shapes are black against the immense cosmic pyramid.

Reaching the small rise, the view takes away our breath.

The higher elevation compared to the surrounding area helps us perceive this immense light as even more evident, going below the geometric horizon.

We feel immersed, even lost in a sky that is alien to us and which now shows us all of its overflowing power.

I have already seen the zodiac light superimposed at the galactic center during the past days, but never from such a panoramic observation point. I have to be sincere: for the first time I feel even intimidated by the impressive demonstration of strength from the Cosmos and now both Marco and I certainly feel smaller than ever.

The intimidation doesn't become anguish, however. It's impossible: all of our molecules are part of this Universe, children of ancient stars, just like those we are observing. They have been here on the Earth for billions of years and have known this sky much earlier than meeting each other to form bodies and minds.

It's instinctive to command and guide our gaze toward an unknown which, however, can't frighten us and, deep down, doesn't even seem so unknown. It isn't; it can't be, because in reality we have been admiring, without a memory of it, this vision for 4.5 billion years.

We strongly feel a profound respect that becomes admiration and has nuances of subjection, aware that, as much as we try, never in the history of man could we have built a view like this.

However, we don't endure passively like witless subjects who bow to the whims of their king. Rather, we feel like an integral part of this vision, because we, although small and insignificant, are one of the billions, perhaps even billions of billions, of consciences with which the Universe wanted to surround itself, to admire its perfect and uncontaminated beauty.

This is the spirit with which we, almost carelessly using our cameras, try to take pictures simply to hang them up in our bedrooms or to keep them on desktop on our computers, with the certain prospect of spending sad days in the future under our luminous bells and hoping that someday we can be reunited once more with our ancient origins.

We forget about ending the exposure, our half-eaten dinners, our friends who are just a few feet below us and who might have needed a hand up. There are just us and the sky; us and the pure ecstasy for at least ten, maybe fifteen minutes. Now, the astronomical evening has officially begun.

Orion is fairly high on the horizon.

Rigel, with its light blue tonality, is more or less at the same height as the Pleiades which, overturned, seem to want to quickly arrive backstage in this theater that doesn't include them as protagonists.

I pause for a moment in this area of the sky, because it appears that I'm seeing an unusual concentration of brightness barely perceivable looking straight ahead and more evident looking sideways.

It's surely light pollution: it would be like being immersed in a city with the street lights aimed upward. It doesn't seem to me to be the case!

This swelling up of the light seems to resemble one of those perfect lakes formed by a tributary which, in this case, comes from the west, and an outflowing river, much fainter but perceptible, which slowly tapers off from the opposite side into the bottom of the sky.

The tributary, looking at it well, clearly seems to join with the zodiac light, which still dominates the western horizon, more evident than ever.

Right now, when I think I've all of the extraordinary phenomena of a dark sky, I have to think again.

This is also one of those phenomena that are read about, with extreme skepticism, only in books on astronomy.

The term is ugly to write as well as pronounce: gegenschein.

What is it? In German, it literally means "reflected glow", but this doesn't help all that much.

I don't know if I'm the best one to explain it, but the gegenschein is an oval-shaped reinforcement of the zodiac light, in the spot exactly opposite the sun.

Once again, the culprit is the dust along the plane of the Solar System.

Near the Sun, it generates the zodiac light. As we watch it at a gradually increasing angular distance, the percentage of light that reaches the miniscule dust particles decreases and, therefore, also its visibility.

In the spot exactly opposite the Sun, however, there is a reinforcement caused by the fact that we are observing the glowing particles in front of us with, therefore, a phase that is always full. It follows that a greater amount of reflected light is responsible for this weak glow in the middle of the sky.

I never saw, or maybe just never noticed, the gegenschein during the preceding evenings, not even under the perfect sky of Chillagoe. This evening, even though I saw fewer stars, the phenomenon is evident, how the zodiac light that crosses the entire ecliptic, is reinforced right here in the zone of the Pleiades and continues on until it embraces the sky from horizon to horizon.

It's truly impressive; I've never seen anything like it!

Now the real hunt for the things I've never seen before begins.

The first objective, suggested by Malù who, in the meantime, has quickly pointed it with his binoculars, is the Crab Nebula.

It's a tiny ball, but is perfectly visible with both the binoculars and a telescope: the elongated form is evident. If it weren't for the slight enlargement, its brightness is comparable to that of an 8-inch telescope under a dark sky over the plains.

Our next objective, suggested by Marco, is the M78, a reflection nebula in Orion.

I remember what the Italian Wikipedia page has to say about it:

"The object can even be seen with 10x50 binoculars, although you need a very clear sky to observe it and to observe it with a telescope with a diameter of 60-70mm (2.3- 2.75 inches)."

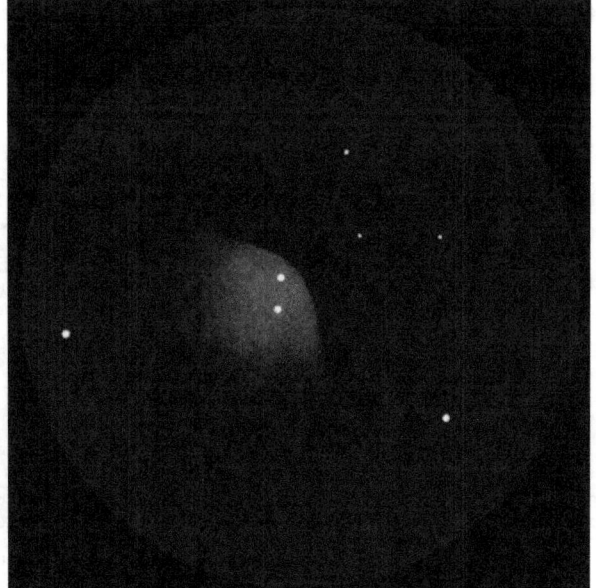

M78. I've never been able to see it from the city, and yet here, it's all too evident with a 32 inches telescope.

This is why I'm a little skeptic.

In fact, Marco tries looking for it, but can't find it.

"You try, Daniele; I'm convinced it can be seen."

He hands me the monocle and I find it in just a few seconds, small, bright, in perfect contrast against the sky, with a shape that brings to mind a cosmic guitar pick: "No, just look how clear it is; it almost looks like a photograph!" I exclaim, passing the baton, first to Malù and then to Marco, pointing out the position with the laser (and there goes another ruined photograph!)

Everyone agrees; even Francesco, who is observing it with the binoculars, through which, thanks to the two-eye view, is even more contrasted.

Marco has sniffed out the potentiality of the sky and ups the stakes. "Daniele, give me the monocle; I'll try to see the Flame Nebula!"

I don't say anything because I know by now that anything is possible. In fact, "Nooo, guys: I can't believe it! You can see the Flame, it's really clear! I can clearly see its shape!"

"Marco, don't say idiocies!" Malù tackles the issue head on.

"Daniele, you look, and then tell me if what I said is true!" Marco echoes.

I take the monocle and focus on the Alnitak zone.

I can't believe my eyes.

Once again, it seems like I have a photograph in front of me.

The nebula is not only visible, although semi-transparent but it also has the typical form of a flame visible in each shot, with evident, slightly darker veining.

I take a better look to be sure, and it's still there: it's not a product of my imagination!

I don't say anything.

I pass the monocle to Malù, who exclaims, "Whoa! It really is there. Spectacular!"

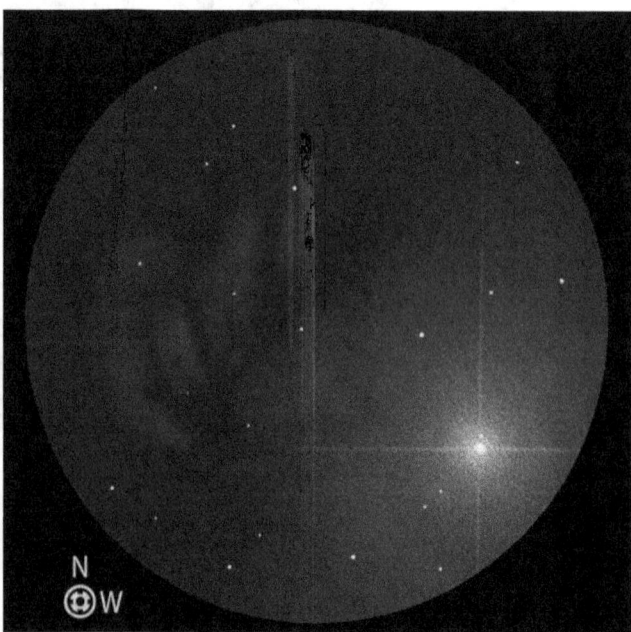

The Flame Nebula seems to be a black and white image. Clear, veined, and even large dimensions with both the telescope and the 70mm (2.75 inches) binoculars!

I look again through my little telescope, but after a few seconds I hear Marco who, having taken over the binoculars, exclaims, "The Rosetta! The Rosetta! It's there, sculpted like in a picture!"

Clamorous laughter, then he continues, "I can't believe it, but where in the world are we! Look at it, Daniele; it's spectacular!"

Well then, after having pointed at the zone with the laser, I can observe, with my pitiful 80mm (3.15 inches) reflector, what I had never been able to see through the years even with a 250mm (10 inches) telescope. The Rosetta Nebula, with its clearly defined, delicate petals, wraps the brilliant open cluster at its core like soft silk.

Now the view reaches absolutely emotional moments.

I contemplate it, even more clearly with the binoculars, and I can finally, perhaps truly for the first time, appreciate my observations of the deep sky.

There's no need for a camera if a small telescope already lets you see the same details through the best instrument of all: the eye. Cameras aren't able to capture the emotions, all those nuances, a much broader field. In 1/20 of a second, the eye sees more than a photo shows in several minutes, without the gift of cosmic rays, sound raining down, burned pixels and interminable hours on the computer. Here, you can touch the Universe with your hand and you can see it live.

What expert observers say regarding this object comes to mind, something that goes like this: *"The Rosetta Nebula is a fairly difficult object for small telescopes. With the best binoculars, it can be identified as a faintly lighted and shapeless aura that surrounds the open cluster."*

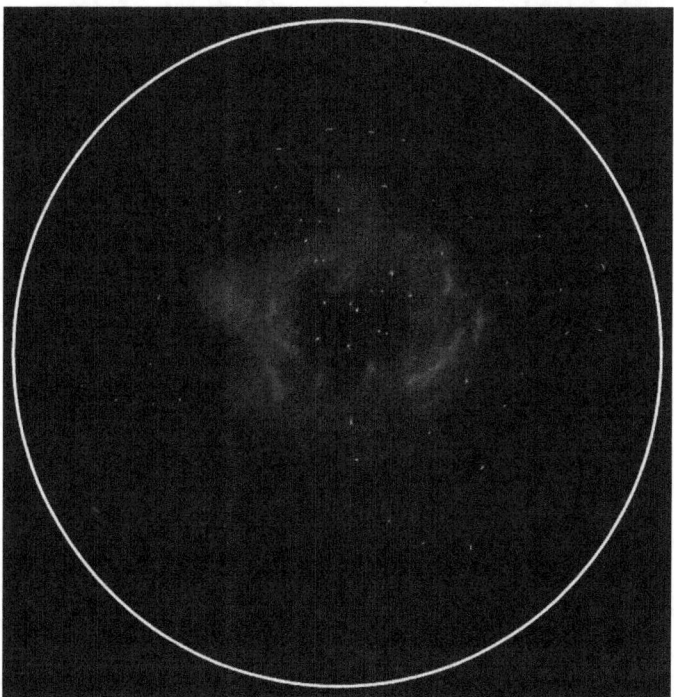

The cosmic rose of the Rosetta Nebula, detailed and sculpted, is identical to the photograph.

This sentence is just wrong, if you have the proper Sky at hand; this cosmic rose has bloomed perfectly in our small instruments.

A bit lower, the M33 galaxy, very easy to point out, reveals a surprisingly brilliant nucleus and a few hints of irregularity. Andromeda, especially with binoculars, extends for several degrees. But these are easy objects.

Marco wants to aim even higher: the North America Nebula, which shines very low on the northwest horizon.

He, with the binoculars and I, with the telescope, begin a very easy search.

Without even the time to doubt, in fact, and it is very clearly there.

It's not the classic glowing, shapeless halo that I observed in the dark (I thought) sky of Forca Canapine last summer; here it has the exact shape that made it famous.

While the northern part of the "American continent" is difficult to trace, the "Gulf of Mexico" is more than evident, to the point that we wonder if it's our imagination at work rather than our eye. But, it's not like that. Expert observers say something like this:

"The North America Nebula extends over a surface area equal to about 10 times the size of the full Moon, but it's brightness is low and can't be seen by the naked eye; it is located about 3° ESE of the bright star Deneb (α Cygnus) in the direction of a very rich and glowing section of the borealis Milky Way. Using binoculars with a wide-field view (of about 3°) it appears as a foggy, arch-shaped spot of light, barely perceptible and only with the condition of sufficiently darkened sky."

It's all true, for a medium dark sky. Instead, the shape is perfectly detectable by the naked eye, while binoculars and my little telescope seem to become gigantic professional instruments that show something that is apparently impossible.

And yet it's here; the Gulf of Mexico is an arched tongue perfectly detached from the background sky. It's all so incredible that I'll never forget it...

How difficult is it?

During my wanderings around Italy, showing the sky with telescopes and speaking about the Universe, I have met many curious people, who are fascinated by astronomy. It's normal; after all, we're talking about the Universe, the everything. And yet, faced by so many curious people who listen to me and ask me questions, only a very few take the next step: exploring and knowing the Universe above us through their own efforts. And yet, it would be so easy; the stars are always up there above our heads; why not want to know them and investigate what's there? Then maybe we won't like it and we get bored, but if we're just a little curious, why not try it?

During the years, the opinion that observing the sky is too complicated has spread. In recent years, all sorts of technology have appeared, promising to make our life increasingly easier but also making us lazier. Because, in fact, there's one thing to which we can't object: thinking is exhausting. Knowing, exploring, and amusing ourselves in such a pure and profound manner is much more exhausting than going to the disco or being suckered in by a few jokers who say idiocies on TV. Being entertained doesn't mean turning off your brain!

Astronomy proposes a much more beautiful and enduring idea for entertainment. When we observe the sky and then return home, something has changed inside us because the sensations we have been given and the sense of satisfaction we receive are real and don't disappear the moment we turn off the TV. And, although there are some who might not believe us, amateur astronomy isn't difficult. The difficult part is beginning, finding the strength to let go of pretend conveniences that have transformed us into beings without the desire to think and dream. Therefore, let's give ourselves a gift: let's begin to dream again, to want to discover, to improve and to have fun doing something that will remain inside us forever.

Everyone can know the sky, from little children to the elderly. You don't need a college degree; you don't need to understand math or any other strange subjects. The amateur astronomer, is just a person with the dream of reaching the stars with his own telescope, who knows he can get what he wants, when he wants and in the way he wants.

Of course, there are things to learn and it won't be easy at first, but we have to face many, much more demanding challenges in life, some of which we have already overcome without even realizing it. We have learned how to walk, to talk, to interact with other people, to live in this world. We have learned how to read, write, ride a bicycle, and maybe even drive a car. We have quickly learned how to use a computer, a mobile phone and to finish a really difficult videogame. In short, if we really want to learn to know the sky, it's less difficult than almost all of the challenges we have faced up to now. It's enough to want it and to have lots of patience. The best way to conquer this new challenge is to not stop and think too much and to try to begin, even with a little healthy recklessness; jumping in and trusting in our sensations and the Universe, which knows how to take care of itself and adequately reward those who choose to explore it with respect.

The Sky seen with the naked eye

You can't begin a long trip, as fascinating as it may be, without first buying your ticket, packing your bags and planning a minimum part of an itinerary.

In the case of observing the sky with a telescope, the situation is very similar. It's difficult to set out in search of the hidden treasures of the Universe if we don't know where they are, if we aren't oriented in the sky or if we don't know how a telescope works. Therefore, a minimum of preparation is indispensable to keep us from being bitterly disappointed later.

How can I be so sure?

Naturally, it's because it's something that I've gone through myself many years ago.

When I received my first telescope for Christmas, after being enchanted by that spectacular view of the lunar craters with those binoculars, I didn't know what to do. How did I aim it at the sky? With what did I observe it? And, above all, what was I looking at? I thought I'd try aiming for the planets: yes, but where are they and how can I distinguish them?

I began by buying a couple of books and a few pamphlets at the newsstand. One evening I saw a beautiful galaxy on one of those pages; its name was Andromeda and it was the brightest and closest. Wow, you could even see it with the naked eye! In fact, it is the furthest objective here that is visible to the naked eye; another galaxy containing hundreds of billions of stars. All this was more than sufficient to excite me and decide to observe it. Where would I find it? Simple: in the early evening northeastward, at thirty degrees from the horizon, 8° more northward from the third star of the constellation with the same name. It should show up like an elongated cotton ball. Simple, no?

It just might have been simple, but I didn't see the Andromeda galaxy, not that evening or in the ones that followed.

I hardly knew where northeast was, I had no idea how far 35° could be and I had never seen the Andromeda constellation.

I learned a fundamental rule the hard way: before using a telescope, first you have to know the sky at least a little, be familiar with distances in degrees (known as Angular Distance), how to find your bearings with cardinal points, know how to recognize, with the help of a map, the most important constellations and know what to expect. At a distance of several years, I'm convinced that my eyes had surely landed on the galaxy more than once during those evenings, but I didn't see it because I was expecting to see something much more striking.

The best gift we can give ourselves is to begin studying the sky a little with the naked eye. Yes, I know, that's probably not what we were expecting. And then, if we already have a telescope, we'll be dying to point it at some object. We can do it, maybe at the Moon (by the way, have any of you noticed that the Moon always shows the same face? No, we'll never see the other half from the Earth!) so we can placate our unbridled desire to skip steps a little and begin to explore; but then, we also have to dedicate a little time to understanding where and how to move around. How long? To know the sky well might even take years, but to begin exploring it with our instrument, just a few evenings and afternoons. Keeping all our commitments – and the fact that every time we decide to watch the clouds someone will do everything they can to ruin our fun – in mind, let's just round it off and choose a number: ten days.

The Andromeda constellation and the small galaxy in the forefront, barely visible to the human eye. I looked for it many times; maybe I even saw it, but I would never have expected to see such an evanescent disc. Who can find the galaxy in the photo?

Ten busy days in which, if we apply ourselves, we will be able to take advantage of our instrument, if we already have it, or to satisfy our desire for the Universe by buying our first telescope and, naturally, finish reading this book.

The first day has probably gone by, so there's no need to do anything else. But if we still have energy and the desire to experiment, let's do it. In fact, this is another golden rule: don't let others, including a book, tell us word for word what to do and how to do it. Advice is welcome, but then, we're the ones who decide.

So, we have finished our "homework" for today and we can do what we want, even point our telescope at the stars and act for ourselves. Let's begin to taste one of the biggest lessons of life that can give us the limitless space above us: freedom, total freedom to choose, to navigate, to rejoice and to make mistakes, but always to learn.

What can be observed with the naked eye?

At the beginning of my passion for astronomy, I thought that the telescope was the only instrument that would let the Universe astonish me. I was wrong, but I only realized it many years later.

Besides the constellations, there are phenomena and views visible by the naked eye that no telescope could ever show us, for one simple reason: we have a complete view of the sky with our eyes, while any instruments can only show us a small portion. We'll discover some surprises directly in the field through our own efforts but in the meantime, I'd like to reveal something that will let you understand how powerful our eyes are. I already spoke of Andromeda, the furthest object from us visible to the naked eye, but there are many more examples.

At the core of one of the most beautiful constellations, Orion, there is a large, glowing nebula, perfectly visible to the naked eye. We'll already have come across it many times but have never understood that that small star – unfocused at times – is the nebula itself. It's not spectacular, but we are observing a blanket of very rarefied gas extending for tens of light years, meaning hundreds of thousands of billions of miles, inside which thousands of stars and hundreds of planets are being born.

We won't have to struggle to see a small indistinct cloud that doesn't seem at all like a star at the core of Sagittarius, a typical figure of summer nights. It's the Laguna Nebula, another oven for stars. In the Perseus constellation, visible for 10 months of the year, there is, towards Cassiopeia, another indistinct cloud. This time, we are observing two star clusters, concentrations of stars that rotate around each other and are spectacular to watch with a telescope.

All in all, the dark sky, if observed with the naked eye, reveals a large quantity of deep-sky objects. Nebulae, star clusters and galaxies are themselves defined as being deep-sky objects or from space. And this is a big surprise and already a nice challenge, because we can have fun looking for all these small, strange clouds and imagine traveling for thousands or millions of light years into endless space. And there's even more, but I don't want to reveal it yet. Our eyes will be filled with wonder by the beautiful surprises that the celestial sphere can show us when we're finally able to observe it far away from the damaging artificial lights; because, if the sky seems to be a dark and dull place, it's due to the only living being on this planet that is able to erase it: man.

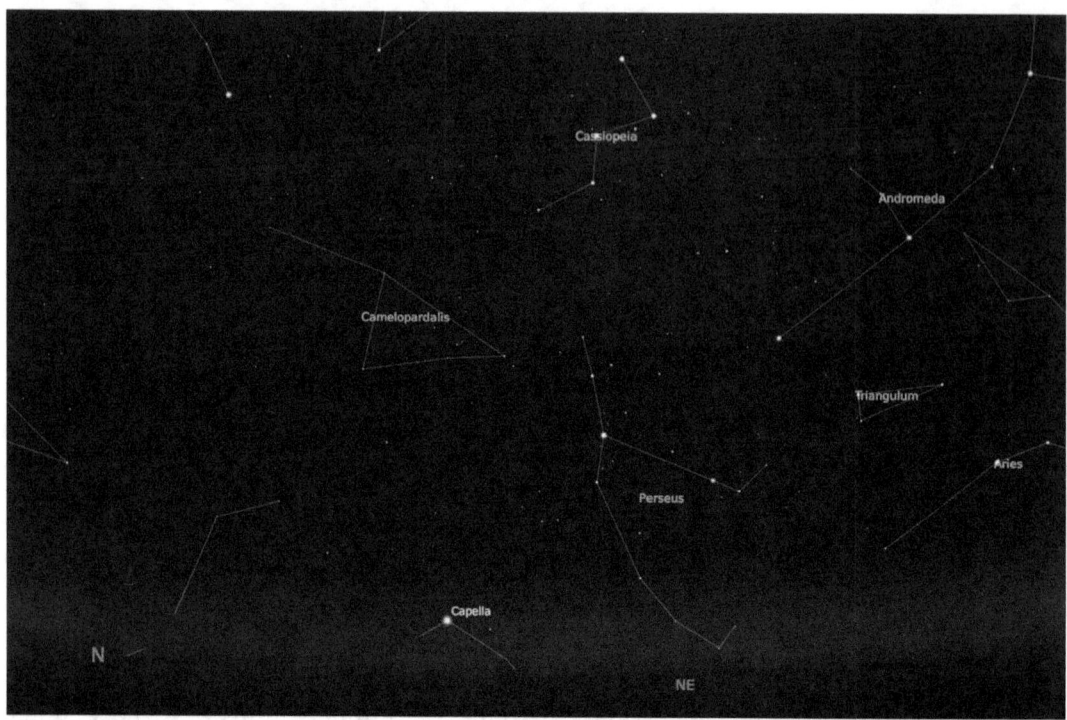

One of the parts of the sky that is most interesting to the naked eye after the summer Milky Way. Among the constellations of Cassiopeia, Perseus and Andromeda we can see at least two indistinct ribbons of light. The first is the Andromeda Galaxy, which we already know; and the second, between Cassiopeia and Perseus, is the famous double cluster, two agglomerates composed of thousands of stars which were all born at the same time, which will be spectacular with binoculars or with a telescope. But in the meantime, let's enjoy the view with our eyes, an extraordinarily powerful instrument.

The light pollution

I was lucky enough to grow up in the country, about 19 plus miles from the closest (small) city, so for many years I always considered the dark sky as something normal, just like the Sun's presence during the day.

Then, growing up and moving to the city I understood, also through talking with other people, that I was extremely lucky because very few people have seen a truly dark sky.

It's called light pollution and, as can be imagined, it's mankind's fault. With the thousands of lights directed toward the sky, the stars have slowly gone out and, at this point, the situation is dramatic. The last uncontaminated sky in Italy was destroyed more than 30 years ago. Since then, no one in my home country never saw a perfect dark sky. Even in the most out-of-the-way countryside, on the peaks of the highest mountains, and even though one might almost reach perfection, there will never be absolute darkness.

In the silence of newspapers and TV news programs, accessories to the insensitivity of the majority of the population, we have been able to eliminate our most precious and enduring asset: the view of the Universe. The night stars have illuminated the uncontaminated nature on this planet for billions of years, while being admired in all of their splendor. Now, on this boot-shaped spit of land embedded in the center of our history, the night sky has been eliminated. It is perhaps the greatest destruction ever done, because we have been trapped in our artificial world like stupid mice, eliminating billions of years of history and evolution, in just a few years.

Light pollution is a silent and apparently harmless form among the many ways that man poisons himself and the entire planet, but it is indeed very serious.

Aside from the danger to nature and animals which are now faced with days that are perpetually illuminated even when the Sun is well below the horizon, light pollution is the number one enemy that we must fight against for our purpose.

But, how damaging can the city lights really be? And how far away do we have to go? For both questions, I have had field experiences which have helped me understand much more than what the enthusiasts who fight against this monster have said throughout the years.

When I bought a nice 10-inch cardboard reflector, I already lived in the suburbs of Perugia, a city that's not terribly large, but nonetheless polluted. It was winter and I couldn't resist the temptation to try out the new instrument that promised to let me see the celestial bodies like I had never seen them before. With my small worn-out guide to the sky, I had pointed a couple of things that should have been in the eyepiece field: the Andromeda Galaxy (which I had finally found a few years earlier!) and the Orion Nebula.

I was already a sky expert and had already observed these objects with my 3.2-inch and 3.6-inch refractors at my grandparents' house. With a 10-inch diameter telescope, I should have been able to see even the colors of the Orion Nebula and the bands of dust in the spirals of Andromeda... Wow! When I aimed at the two objects, I received an incredible disappointment. From that city sky where I couldn't even see the entire Little Dipper, the Orion Nebula and the Andromeda Galaxy seemed even weaker than the observations done with much less powerful telescopes under dark skies. I even thought that the instrument didn't work, but when I decided to take it, with great effort, to the countryside, I understood that I had underestimated the power of light pollution and the doors to the Universe were literally opened to me. But this is another story that I'll tell you further ahead, because now I have to answer the second question: how far do we have to go to not be disturbed by the lights?

Light pollution is based on a very simple principle, the same one that lets us see the sky clearly during the day. When intense light goes through the air that we breathe, the latter isn't perfectly transparent but acts like a fogbank, intercepting part of the light and reflecting it in every direction.

The disorderly reflection of the light is called diffusion and this is the phenomenon that makes the sky appear blue during the day. If we were in space, outside of our atmosphere, the sky would appear completely black, even with the sun high on the horizon, because there wouldn't be particles of air to diffuse the light.

With artificial lights, thousands of times more powerful than the brightest star, something identical happens. So, only a hundred stars can be counted in the big cities.

When I went to New York the first time, I could only see Jupiter in the bright-as-day Manhattan night sky. Unfortunately, however, it wasn't enough to just leave the center of the lights to see the sky better.

Misfortune would have it, at least in these cases, that our atmosphere is extended by at least 6 miles higher; therefore, the particles of air that can diffuse the artificial light and clear the sky are found even at elevated heights. The result? Unfortunately, disturbing. Looking south from a dark sky like the one observed on Mount Amiata in Tuscany, you can easily see the glow created by the lights in Rome, more than 63 miles away as the crow flies. You can distinctly see the annoying glow of the lights of L'Aquila from the skies of Forca Canapine, a locality well-known by enthusiasts of central Italy, on the border between Umbria and the Marche, more than 37 miles away.

So, not directly seeing the lampposts isn't enough but, as a matter of principle, we should move at least 125 miles away from the big cities so that we don't even see the entire column of atmosphere above the diffuses the light from the surface light.

One hundred twenty-five miles without a city in such a densely populated State like Italy doesn't exist and this explains why even the most isolated places will never have a perfect sky. Naturally, this doesn't mean it will be impossible for us to have beautiful observations. Between observing the sky from under the lampposts in the center of a city and from the middle of a desert, which doesn't exist in Italy, there's a good compromise in the middle, which can be summarized as follows: if we want a good sky we have to go at least 30 miles away from a big city. Twenty is probably far enough away from a medium-sized city.

The best skies are found at high altitudes, because the higher up you go, the stratum of air that the stars must cross is thinner and there is less absorbed light. The worst places are in the polluted valleys, a territories that are often affected by high humidity, which is as much an enemy of the faint starlight as is light pollution.

The differences between a dark sky, far from the city (on the left) and a sky illuminated by artificial lights (on the right). Light pollution is extremely harmful in observing the sky, except for observing the planets and the Moon. Credits: Jerry Lodriguss

What are the constellations?

Now it's time for action and beginning to understand something about this beautiful sky. Where do we begin? With the constellations, of course, a word that we have often heard mentioned. And just as frequently, we'll have heard at least a few solemn names: Orion, Big Dipper, Scorpius... do they us anything?

The constellations are imaginary designs that the ancients, mainly the Greeks, have imagined that they've seen in the sky. Certainly, dismissed this way, the definition seems to lose much of the romanticism and the solemnity of the sky and doesn't do justice to the philosophical and spiritual side behind them.

To understand what the constellations are and, above all, their meaning, we have to think of ourselves as citizens of times long ago, when electricity didn't exist and therefore, even less so mobile phones, TV, radio, internet or computers. A world that is still in the hands of Nature, which decided the rhythms and a large part of the life of that poor population that was still at peace with the planet that it called home.

The skies were perpetually dark and clear thanks to the absence of any source of light and pollution, a word that would be created many centuries later.

After having satisfied their basic needs with food, there wasn't much else to do in a world that was frightening at times because it was rich with inexplicable phenomena. The sky, that array of stars that appeared every time the Sun went down, was both spectacular and terrorizing to all men, and was certainly the only amusement available to the entire population.

The indescribable power of those tiny flames that don't seem to be randomly placed couldn't be anything but the fruit of superior minds that commanded the world and human events: the gods. So then, the ancients saw the most evident manifestations of the presence of the gods – as both protectors and severe judges of humanity as a whole – in the firmament. Those stars, therefore, had to be flames hung from a dark sphere that came to find them every time that Apollo, god of the Sun, finished the flight in his flying chariot while pulling the fiery ball that brightened the day. Little flames that weren't randomly placed, but which seemed to tell myths and legends which involved gods and human beings, so different in appearance but extremely similar in substance.

Therefore, with the stars of the celestial sphere, the ancients used something that in our days is beginning to become a rare commodity: their imagination. They invented stories, figures, myths and legends and tried to make sense, in their way, of something that still, for thousands of years, wouldn't be possible to explain even with the power of science.

At a distance of many centuries we have understood that those things above our heads aren't flames that seem to be hung from a gigantic dome, but stars that are much further away than any imagination, that they all seem to be the same distance away only because our eyes aren't able to see into the depths of space. They have nothing to do with the gods, neither in the past nor in the present and the designs that the ancients have sent down to us are only imaginary figures that don't correspond to reality. However, the constellations are useful. They're useful because they narrate an important piece of our history and they are very useful for whoever wants to begin exploring the Universe, because they represent an excellent instrument for finding one's bearings in what is called the celestial sphere. Therefore, the constellations are our referral points for moving among the wide roads of the Cosmo and find hidden objects and treasures.

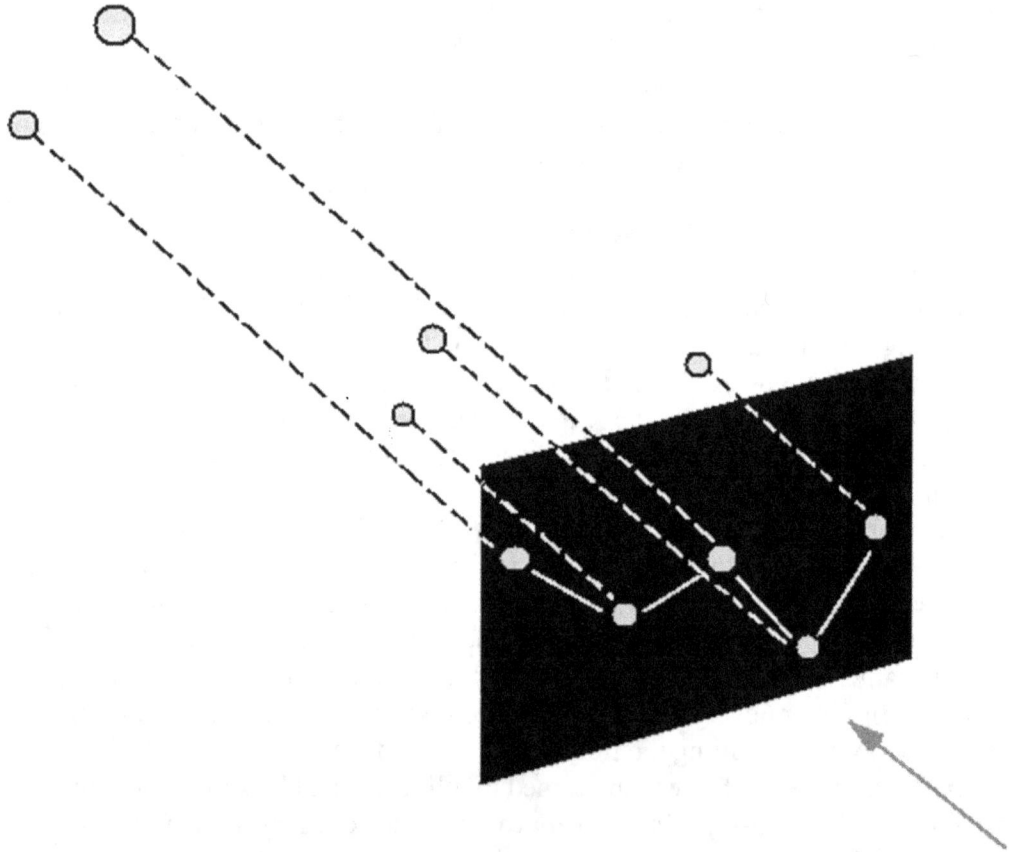

The constellations are imaginary designs decided upon by men. In reality, the stars of a constellation have nothing in common, beginning with the distance between them.

A special star: the North Star

On a calm moonless night, far from the city lights, let's imagine that we are experienced scientists who are trying to understand how the sky functions, how and if it changes and whether it's possible to predict its movements. Practical experience is the best teacher that we can find, even if it might not seem to give us much satisfaction.

Without knowing the sky and without even possessing a compass, we begin writing down the positions of some bright stars in respect to our horizon, with the help of trees, hills and houses. We wait about ten minutes and then write them down again, still on the same sheet of paper but with different colors. We wait a half hour and repeat this operation again with another color. If we feel like it, we can repeat the operation after another half hour.

If we've been good and have chosen stars spread over the sky, we'll already be able to understand many things.

In fact, our drawing will show something strange. Almost all the stars have followed an arc, the length of which changes according to their position in the sky.

In one particular zone, the stars draw increasingly tighter arcs, until they arrive at a rather isolated little star that has remained inexplicably still. That is the famous North or Pole Star. It isn't the brightest, nor is it the first to appear after sunset, and yet it is a very important star that has saved the lives of thousands of sailors before the invention of the compass, and has let humanity push its way into unexplored regions here on our planet. The North Star, in fact, is the only visible star in our latitude that always stays more or less in the same position in the sky.

Almost all of the others follow paths that are similar to the Sun's. They rise in the east, reach their highest point on the southern horizon and then set in the west.

The North Star seems different. Why? What does it have that is particular? And why do all the others move?

The time has come to clarify a few things about the movements in the sky and take the first important step toward knowledge of the celestial sphere and the future observation on the telescope.

The movements of the stars of which we have taken note return every night and the described arcs form a complete circle in 24 hours, exactly the time used by the Earth to complete its rotation. What seems to be a coincidence isn't in reality, because the movements of the stars, including the Moon and the Sun are primarily due to the Earth's rotation. The stars remain fixed in their place in the course of the night; we are the ones who continuously changing orientation because of the Earth's rotation, even though we don't notice it if we aren't looking at the sky.

So, why does the North Star seem to stay still?

The sky's movements exactly mirror the Earth's movements: its rotation in this case. Our planet turns around on an axis, called the rotational axis, which passes through the North Pole and the South Pole. The entire planet rotates around these two points, which remain still. Here on the Earth's surface, we see these movements reflected in the sky, including the way in which they take place. The North Star stays put because he Earth's rotational axis points in that direction of the sky.

It's a completely terrestrial property, because outside of our very little and welcoming planet is a star like all the others that doesn't know that it is so special for those minuscule and fragile human beings.

What we have tried to see with our eyes can be seen much better through a long-term exposure. All stars complete circles – the centers of which are found very near to the North Star – going from east to west.

The Sky that changes

Another of the Earth's movements that we see well reflected in the sky, even though much slower than its rotation, is produced by its orbit around the Sun.

Let's go back to the experiment we did to understand how the sky rotates during the night but this time, let's write down the positions of the stars with regards to the horizon, for at least 3 evenings in the arc of a month, always at the same time.

We'll quickly notice that the entire sky again seems to move from east to west around the North Star, but this time it's not caused by the Earth's rotation, because we have always observed them at the same time.

The path followed by our planet around the Sun, called orbit by astronomers, makes all of the constellations move with the passing of time, excepting, naturally, the North Star.

Every day a star reaches the same point in the sky 4 minutes earlier than the preceding night. It doesn't seem like much, but if we observe the sky two months later, it's more than enough to let us see completely different constellations. This change in the sky, which is completed in exactly one year, determines which portions of the Universe we can observe according to the seasons. In the summer, for example, the Milky Way and the bright figures of Scorpius, Sagittarius, Cygnus, Lyra, and Aquila are very evident. Instead, these constellations are prospectively closer to the Sun, still in the early evening and so they can't be observed unless you are in space where the sky is black even during the day. On the other hand, other portions which were difficult to see before, like Orion, Sirius, Gemini and Taurus, are now visible.

To get our bearings in the sky correctly, we have to do a little practice and understand what we can observe in a specific time of year. A good way to understand it would be to download a few programs from internet, like Stellarium and SkyChart, which simulate the starry sky. From now on, these will be our best allies, to be consulted even when we have finally practiced with our own first telescope.

Wanting to be precise, there are many more movements in the celestial sphere. The truth is that nothing in the Universe remains unchanged; everything changes, even though it's often with times that are too long for our brief lives.

During the passing of the years all of the constellations seem to move in the sky in a particular manner. These movements also regard the Earth and its axis in particularly, which doesn't always remain aimed toward the same zone in the sky. The consequence is astonishing: the North Star is close to the celestial North Pole only in this point of history. At the time of the Egyptians and the early civilizations of the Earth, which appeared about 10 thousand years ago, the current North Star was very far from the north and was therefore completely useless to nocturnal navigators, who probably used Thuban, a star in the Draco constellation, to find their bearings at night. Even though perhaps no one will use it anymore, in the year 14,000 A.D. the new North Star will be Vega because it will be near the new direction in which the rotational axis of the Earth will point. It's a little troubling to think that our North Star, the one that has guided many intrepid navigators over hostile seas is only a small parenthesis in the long walk of our civilization and the history of the Earth. The momentum that will change the North Star and very slowly rotate the entire sky is called precession and is completed every 26,000 years. And so, in 26,000 years our descendants will be able to once again look at the North Star with our same interest.

Looking at the bigger picture and over an even greater time, it's even stranger to think that the sky that the dinosaurs admired, about 70 million years ago, was probably very different from the current sky. Some stars weren't born, others have exploded and all of them have extremely different positions. The constellations that we know now are designs that didn't exist yet and in some ten thousand years will already be modified enough that they will stimulate our descendants to plot out other unrecognizable figures. Our lives, for the Universe, are so brief that it's as though we had the possibility of seeing a gigantic photograph, one or two photograms of a movie that continues on without stopping for 13.7 billion years.

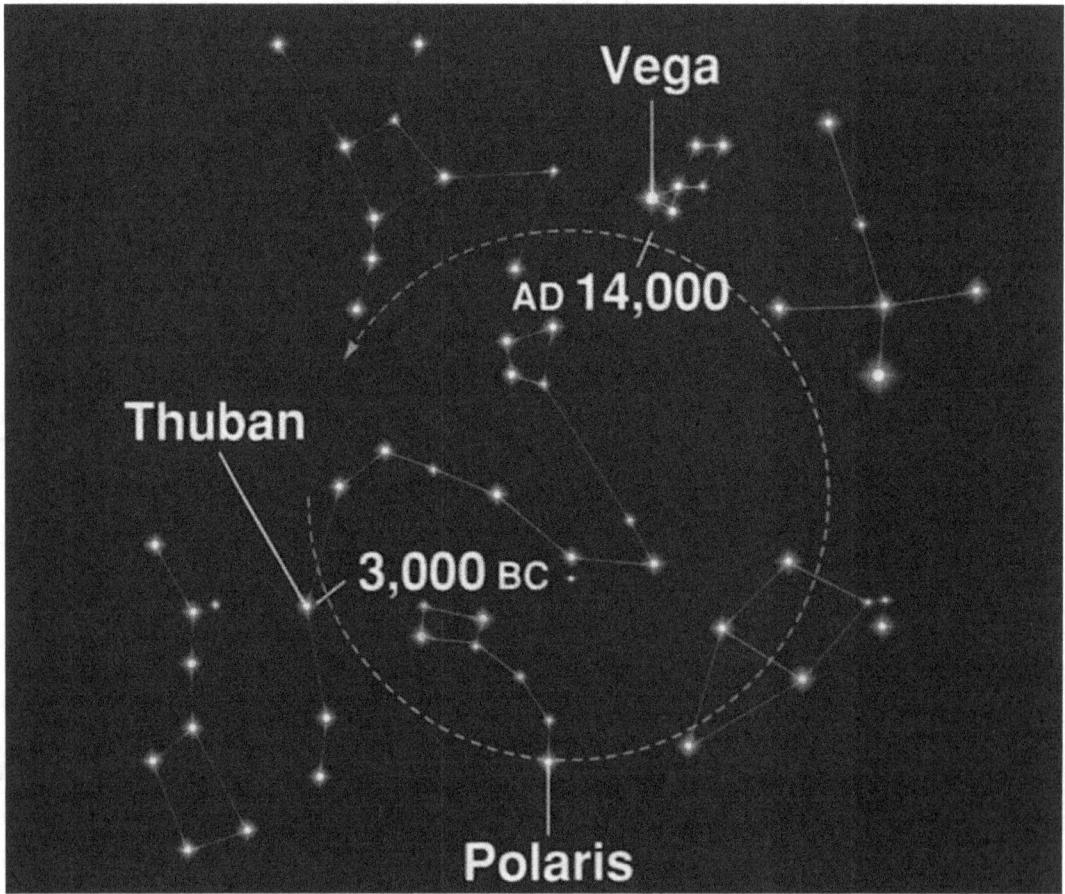

Because of the precession phenomenon, the North Star is just passing through near the celestial North Pole. But in reality, the entire sky is inexorably and irreversibly changing.

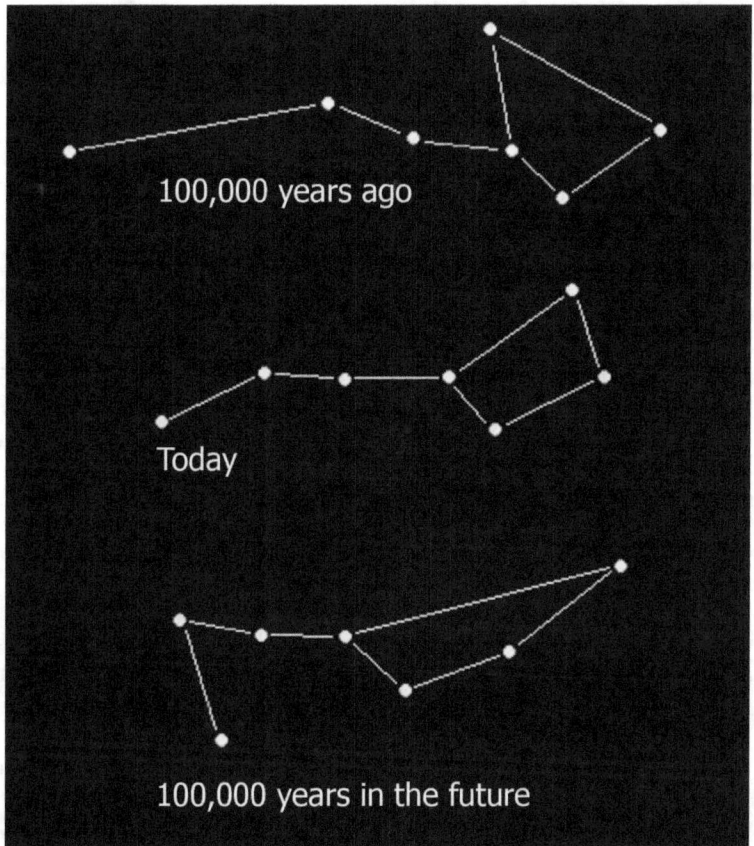

The Universe that we observe now is only a photogram of an almost 14 billion year-long movie. Even the position of the Stars and our familiar constellations are in perpetual evolution, as this simulation of the Big Dipper bears witness.

Let's get our bearings in the sky

By now the constellations are only a way for us to get our bearings in the sky, a little like the streets of a big city. But is this enough to be able to move about in the biggest metropolis we could ever encounter? Unfortunately, no; but we're almost there.

If we put all the pieces together, we can already have great satisfaction from our first waited for and knowledgeable night under the stars.

It's easy to get lost in a starry sky and there aren't any sat-navs that can help us. Fortunately, we're not in any hurry: we don't have to arrive on time for an appointment to avoid being yelled at for being late. Patience is therefore our best ally; let's relax and try to see all this as entertaining, maybe as a challenge to do with a few friends.

The first thing we have to do under the sky is to find our bearings; not with the stars, but with our horizon. A small compass will surely be a useful ally in this phase because it will immediately tell us where to find the north. Let's turn our faces toward this cardinal point and, without our compass anymore, we already know more things than we think. If the north is indeed in front of us, the south is exactly behind us. The straight line that crosses the sky from the northern horizon to the southern passes right over our head at a point that astronomer's call zenith. Putting definitions aside, as we're not that interested in them, all the objects that pass in the south during their journey around the North Star are on this line, at the highest point on the horizon. Let's keep this particular position in mind because we'll need it when we go hunting for objects with our telescope.

Instead, let's return to finding our bearings. Although finding north and south is easy, trying to find east and west is a little bit less so. In this case, we have to remember this phrase: **facing north, east will always be to our right and west is to our left.**

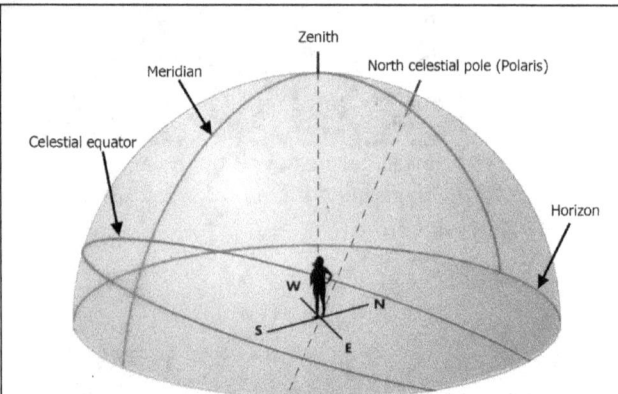

The meridian is an imaginary line that unites the north to the south and passes over our heads (zenith). It is the point at which an object reaches the maximum point on our horizon.

We're good to go because now we know how the sky above us moves. The constellations towards the east, or to the right, reach the highest point along the imaginary line that unites the north to the south and then set toward the west. The precise point where they rise and set depends on how far they are from the North Star because, and it's best not to forget this, the entire sky rotates from east to west around this point.

Right, this point… How do we find it?

Let's find the North Star

The North Star indicates the north, but exactly where is it? How can we immediately understand how to pick it out without having to wait hours to see the rotation of the stars?

When I was little, someone told me – I don't know who – that the North Star was the brightest in the sky and that it would be really easy to identify it. What a joke; it was all a bunch of lies! The North Star is the fifty-ninth brightest star in the sky (!) and picking it out without a precise reference point is rather difficult.

We have two reference points in our adventure. The first is our compass and some notions about celestial mechanics (what a strange words!). In fact, the North Star is at a point on the horizon equal to the latitude of the place where we are observing. If we are watching the sky from a latitude of about 42°, the North Star will be found to the north (look at your compass!) at a height of 42° on the horizon. Easy, right? The problem is understanding how much 42° are. We'll see shortly; let's be content with the fact that the North Star for us at mid latitudes is found halfway between the vertical over our heads (zenith) and the northern horizon, in a zone with very few shining stars. This way, we'll probably see it, but we won't be sure that we've really recognized it; so then, we'll bring out the winning card, a reference that certainly won't let us make a mistake: the Big Dipper.

This famous celestial figure is part of the Ursa Major constellation and is very easy to identify, still northward. The peculiarity of the Big Dipper is that being near the North Star it never sets below the horizon at any time of year. In its rotation from east to west, it is always visible. Also other constellations, not too far away from the North Star, never set and are called circumpolar for this reason. We'll talk about these a little later on because they'll help us find others, according to the time of year.

The Big Dipper can be observed even in the illuminated cities and covers a large portion of the sky: it's impossible to make a mistake. If we watch during the early evening, around 9:00-10:00, here are a few indications on where to find it.

In winter, it's placed almost straight up toward northeast and slowly climbs up the sky during the night. In the spring, we can find it almost perfectly over our head, upside-down. In the summer it's once again vertical, headed toward the northwestern horizon and in autumn, it's extended low on the northern horizon. If we can find the Big Dipper, the North Star is close at hand. The last two stars, in fact, have a big particularity: they point almost perfectly toward our objective, at all times of the year.

Just draw an imaginary line from the star farthest away from the north to the closest one, count the distance between the two, five times and encounter a small anonymous star that is all alone in that portion of the sky. Mission accomplished! Now we can relax and enjoy a little more theory.

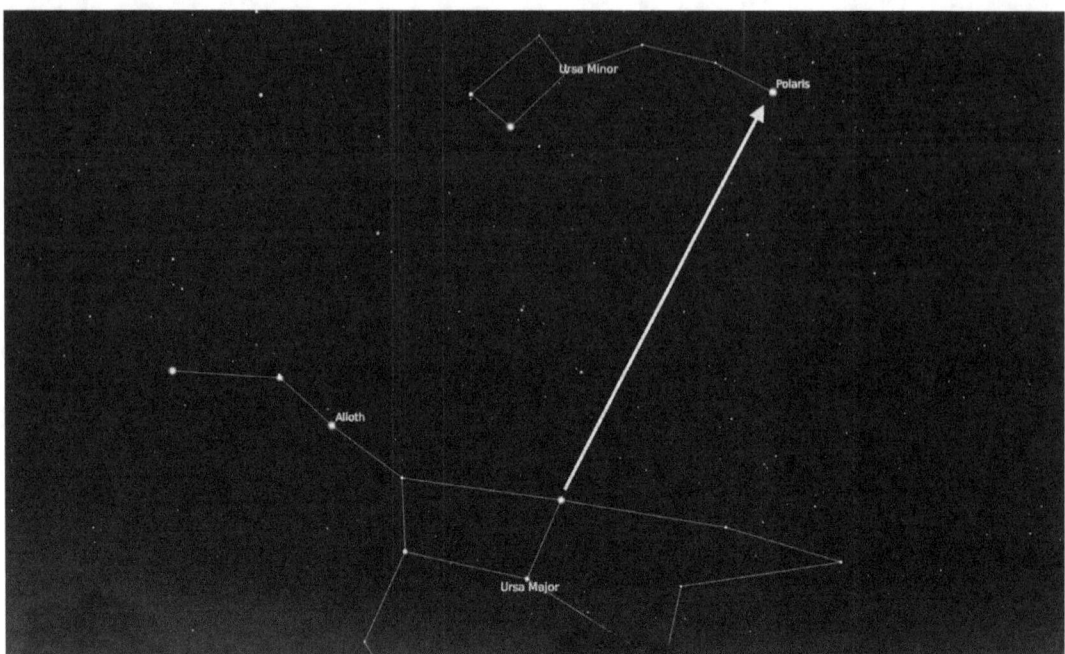

To find the North Star, it's best to identify the Big Dipper and draw a line that starts from the last two stars, Merak and Dubhe and count 5 times the distance between them until you encounter a rather isolated star. That is the North Star. Depending on the season, the Big Dipper changes its orientation in the sky, but those two stars always point toward the North Star.

The Big Dipper rotates around the North Star so, depending on the season, it will have a different position. Here, you can observe it in spring. To the right, a bright star called Arcturus will be the brightest star in the season.

The measurements of the sky

The stars, the planets and the far away galaxies are placed at extremely diverse distances between them and in our regards and yet at night it seems to us that they are all hung on the walls of this dome that hangs over us with its beauty. If we were in a city and had to find a place, we'd use feet or miles, but our common units of measure don't work. Here, astronomers and enthusiasts of all types make use of what are called angular or apparent measures; we no longer speak of miles, but of degrees. A little while ago we mentioned them when I said that the North Star was at a height of about 42°.

Degrees are the unit of measure of the distance between the objects in the sky and it's very convenient for finding your bearings and letting your friends where to find the star you're going to observe.

But how big is a degree? How is it measured? We measure the angles drawn on a sheet of paper with a goniometer, but this instrument becomes impractical in the sky. If we are disposed to give up a little precision, the best way to estimate the measurements of the sky is with our hands. Yes, with our hands.

So then, here are some indications that can be very useful for us:
- Our open hand, extended in front of us covers about 20° of sky between the thumb and little finger;
- A closed fist in front underpins, between the knuckles of the index finger and the little finger, about 8°;
- The distance between the knuckles of the index and middle fingers of that same fist underlies about 3°;
- The knuckle of our distended little finger covers about 1°;
- The diameter of the full Moon and the Sun is about half a degree.

They won't be very precise instruments and we won't look good while we try to evaluate the angular distance between the two stars, but this way we'll have a very powerful instrument within hand's reach and we'll finally be almost ready for our first informed field trip under the sky.

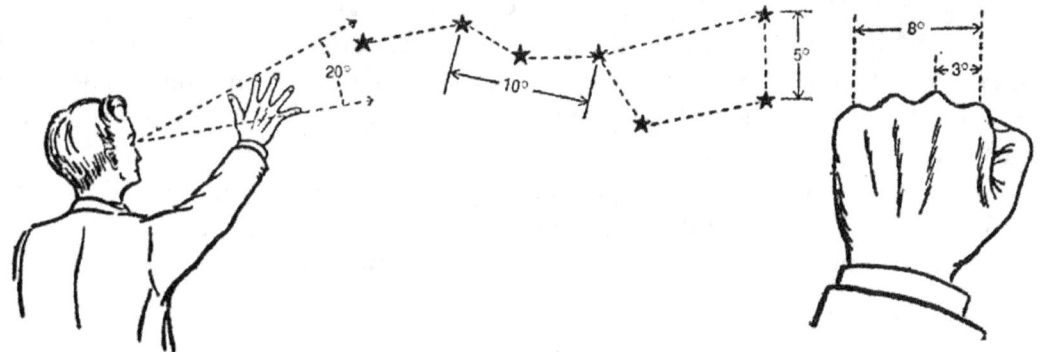

A few simple methods for estimating proportions in the sky.

The magnitude

Another very useful way to get one's bearings in the sky examines the light emitted by objects. In fact, not all stars, and not even planets, galaxies, nebulae and clusters, show themselves to us with the same brightness. There are two principle reasons: they have different distances from us, and the further away they are, the fainter they usually appear; and because they emit different amounts of light, just like our light bulbs are available in different power.

The brightness of celestial objects, for some unknown reason, is measured on a slightly peculiar scale called magnitude. We're not interested in knowing its origins, but only its practical side.

Well, the first thing to know is that it is an inverse scale. The brighter stars have low magnitude values, the fainter ones have higher values.

The North Star, for example, has a magnitude of around 2. The brightest stars can have even stranger values. Vega, a star right above our heads in the summer, shines with a magnitude of 0 (yes, zero!). Sirius, the brightest star in our sky, even has a negative value, equal to -1.46!

The faintest stars that a perfect eye can observe under an uncontaminated sky have magnitudes of 6. 5, maybe 7. These are values that are reached with patience, training the eye to observe objects that are truly very faint.

Those of us who are having our first experiences, can hope to observe magnitude 6 stars from our countryside skies. They seem low values, but numbers don't always have to be big to describe something enormous. There's a 2.5 times difference in brightness between one number and the next. This means that a star with a magnitude 3 is 2.5 times fainter than one with a magnitude 2. Between Vega's brightness and that of the faintest star that we can see the first times, with a magnitude 6, there's a difference of 250 times!

The brightest objects of the night sky are surely some planets and the Moon, which, during the full phase, shines at a -12.6 magnitude. Naturally, the undisputed king is the Sun, which, with its value of -26.85 is a good 500 thousand times brighter than the Moon and millions of times brighter than the brightest star.

The magnitude helps us find our bearings in the sky and correctly read maps and guides. What should we expect if we have to find a nebula at 1° from a third magnitude star? And if there are others in the vicinity with varying brightness, how can we tell which one is the one that interests us?

Once again we'll use the Big Dipper and the brightness of its stars to calibrate our eye. The following map should, therefore, be printed and carried with us during our first official outing among the stars, which is finally almost here.

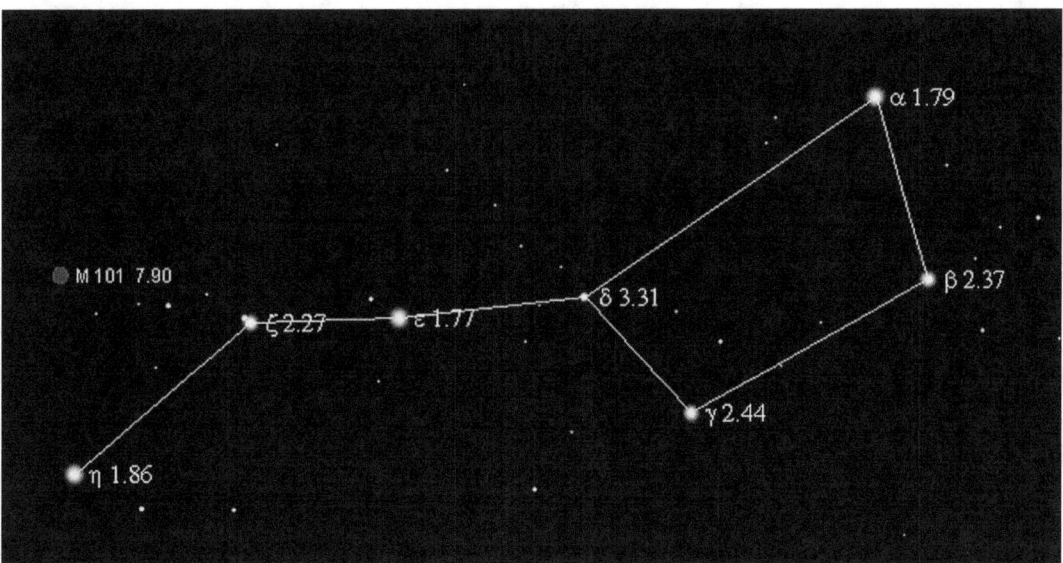

Names and magnitudes of the stars in the Big Dipper, which are very useful as a reference for evaluating the brightness of the heavenly bodies.

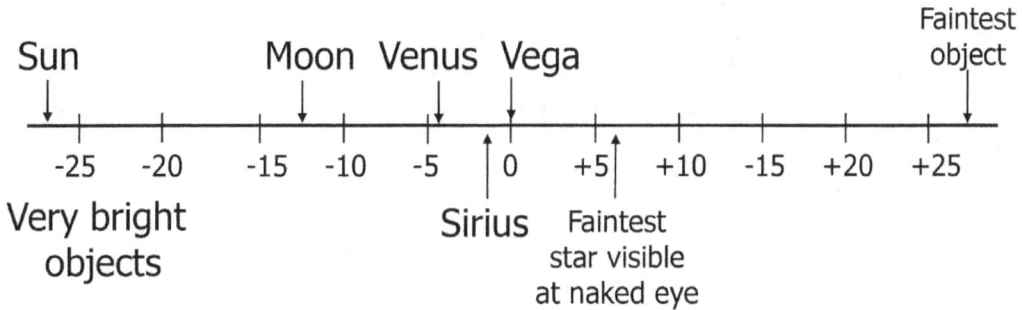

The scale of magnitudes extends from -26 to +30, covering all of the objects in the sky that are visible with current technology.

The first date with the stars

We aren't ready to recognize the constellations and the planets yet, but why not take advantage of a calm night and begin to put what we've learned into practice? Even the Moon would work, since all that we have to do is find our bearings and begin to calibrate our eyes after all the theory we've read.

We can do our "homework" in about half an hour and then we can decide whether to go back home or not.

What do we need to start getting familiar with the sky? First of all, a flashlight with a red bulb. For the eye to see its best at night, it shouldn't be exposed to any artificial light. The dark adaptation, as it is called, to absolute dark is reached after 15-20 minutes under the starry sky. All it takes is a normal flashlight (or a street lamp) to make us lose our maximum sensitivity. The difference, even if we haven't ever noticed it, is strong. The red light, instead, doesn't ruin the adaptation to the dark; this is why all of the amateur astronomers use one!

Before going out, we need a plan: what are going to do? And with what? Programming the first evening will also help us gain confidence with the following nights. All of them, in fact, should be prepared, above all if we have to leave our home. What happens if we forget a map or an accessory? More than once, I've had to go back, and frequently, because of laziness and frustration, I've just stayed home.

So, here is a possible outline for our evening.

What do we want to do? Find the cardinal points, the Big Dipper, the North Star, understand how the celestial sphere rotates, how to calibrate the magnitude scale and learn to measure angular distances. Well, what else do we need besides the red light? A compass, for certain, even a cheap one found in a box of Cracker Jack, as long as it works. Then, it wouldn't hurt to have the maps that we saw in the preceding pages, the ones regarding apparent measures, the position and shape of the Big Dipper, and the ones that measure the magnitude of the stars, also.

Personally, I would also bring a comfortable chair or, even better, a lawn chair, a blanket, some water, a few snacks to munch on, a friend or my girlfriend (or boyfriend), because seeing the stars should be fun, not painful.

I don't have anything else to say (on the contrary, I've said too much), except to wish you a lot of fun and to remind you that you should never be in a hurry with astronomy, since the sky is always up there.

Let's recognize the easiest constellations

I hope that your first knowledgeable outing under the sky went well. Now that we've come back to studying, it's time to evolve and learn a little more about what is in the sky. We're ready to hunt for the constellations, beginning with the simpler ones and going on until we reach the most difficult ones (some of which not even I know!)

Our attention always has to begin from the north, from that zone around the North Star that never sets for our latitudes.

We have already seen the Big Dipper which, to be a stickler, isn't a constellation but what is called an asterism. Setting aside definitions, let's keep the Dipper well in mind, because from there, we will shortly find many other celestial figures.

Before taking off, however, let's concentrate our attention on the part that is exactly across from the Dipper, at a similar distance from the North Star. A fairly bright group of stars that has surely caught our attention, even though we don't know how to connect the dots yet, according to the designs that the ancient people saw there.

As a child, I saw a sort of square hanging from a thread; instead, the Greeks saw the young and beautiful queen of the Ethiopian kingdom, named Cassiopeia.

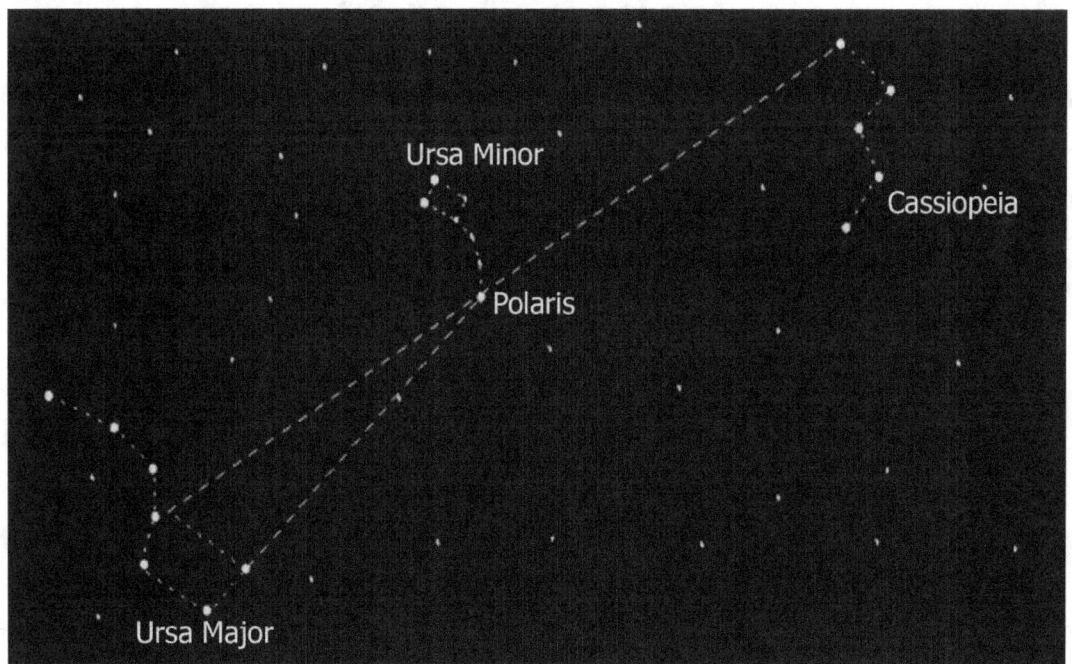

Cassiopeia is found on the side directly opposite the Big Dipper.

The constellation of Cassiopeia doesn't look like the figure of a woman to modern man, but more or less like a much less romantic letter of the alphabet. Depending on the season, these stars form the classic M or W (but I continue to see a square!).

Cassiopeia is one of the easiest constellations to admire, above all in the autumn and winter nights.

There are other constellations in the circumpolar zone which aren't, however, as easily identified. Some, like the Lynx, owe their name to the difficulty of finding them (you need to have sight like a lynx or an eagle-eye!), so we don't need to recognize them right away.

The Little Dipper is between Cassiopeia and the Big Dipper. Its point is identified exactly by the North Star. Contrary to its "big brother", it's difficult to identify because it's composed of very faint stars. If we are able to see it, it means that the sky we're observing isn't at all bad. So there, without even thinking about it, we have found a quick way to measure the quality of the sky. We are just too good!

Now, the moment has arrived to start hunting for the other constellations in the sky, but before doing so, we have to know which ones are visible during the season in which we have decided to observe them. Since I'm writing these pages, I can't know, so I the following pages there is a quick look of the most beautiful constellations grouped according to season, to be observed in the early evening, around 9:00 pm.

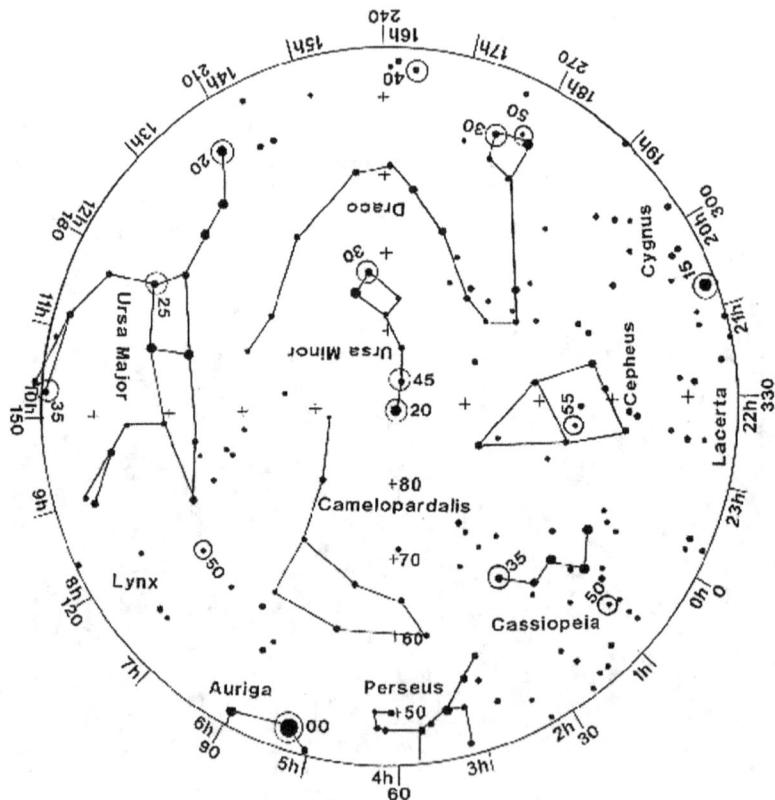

Map of the circumpolar constellations visible from the mid northern latitudes. With the exception of Cassiopeia and the Big Dipper, the others aren't very bright. Facing north, this map shows their position at 11:00 pm during the early part of July. Every time we rotate the map by 90° counterclockwise, we'll see the position of the circumpolar constellations three months after the preceding date. Take note of the Latin names, which are the standard used around the world.

The sky in spring

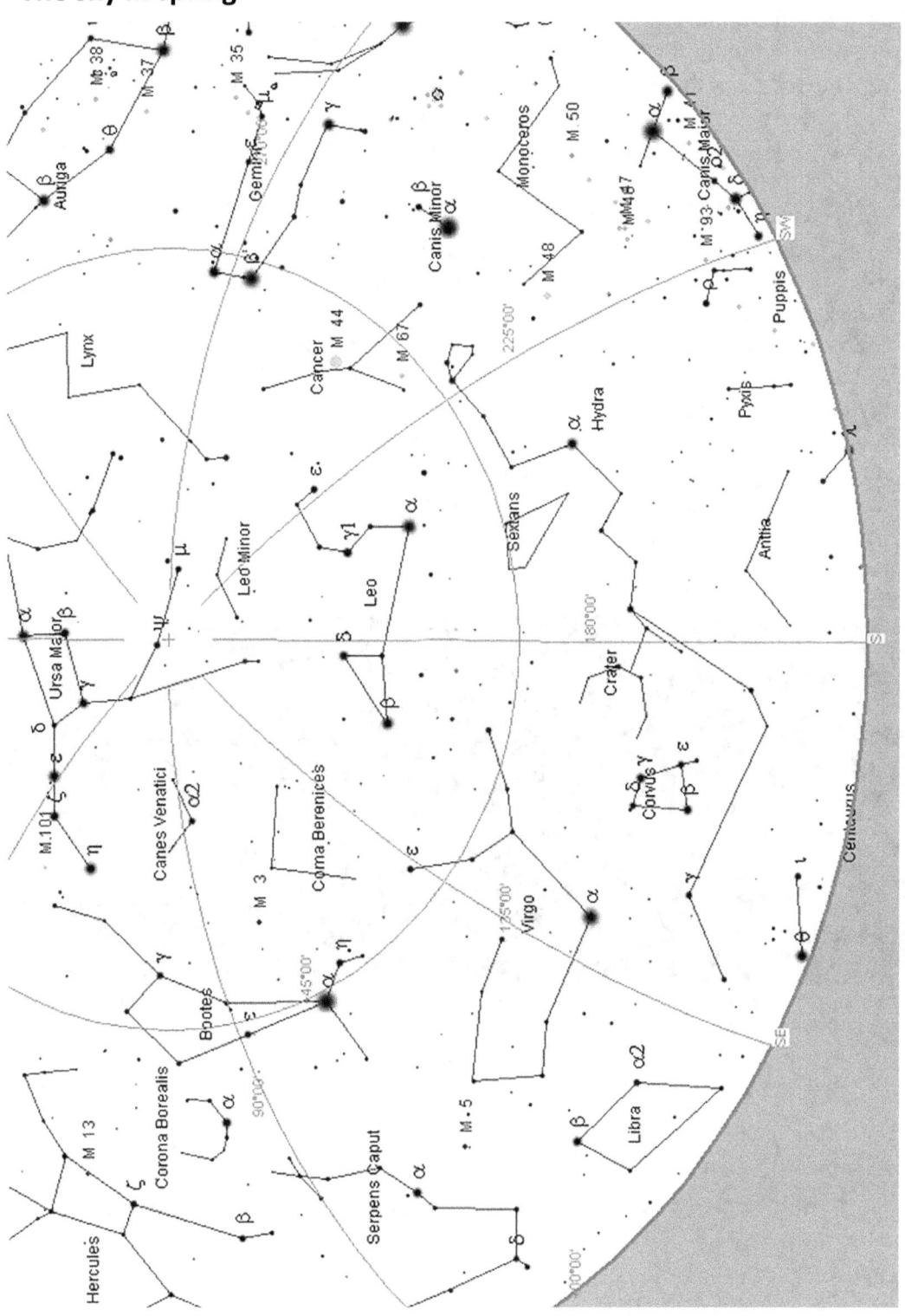

In spring, we can identify many other constellations, still taking off from the Big Dipper, which is almost directly overhead. Extending the handle of the dipper along the arc formed by the stars ε, ξ and η, we run into a fairly bright orange star: this is *Arcturus*, a red giant that is part of the Shepherd Constellation (*Bootes* in Latin), from the characteristic kite shape. Still following the southward arc, we run into another bright, although not as bright as *Arcturus*, white-colored star: this is *Spica*, from the Virgo constellation.

If we go back toward the Big Dipper and consider the extension of the stars δ and γ, or from the two that indicate the North Star to us, (α and β), and we go southward, we run into a brilliant star: *Regulus*, from the Leo constellation, easily identified by the brightness of its components and its unmistakable shape.

There is a low concentration of stars in the spring and this allows us to identify the constellations that are present with greater ease. This is due to the fact that we are observing a part of the sky that is perpendicular to the galactic disc, which projects toward deep space. It's no coincidence that spring is the best season, together with autumn, for observing the galaxies. Winter and summer, instead, are the best seasons for nebulae and star clusters, or rather, for all galactic objects.

The maps found on this and the following pages are surely more useful than any explanation.

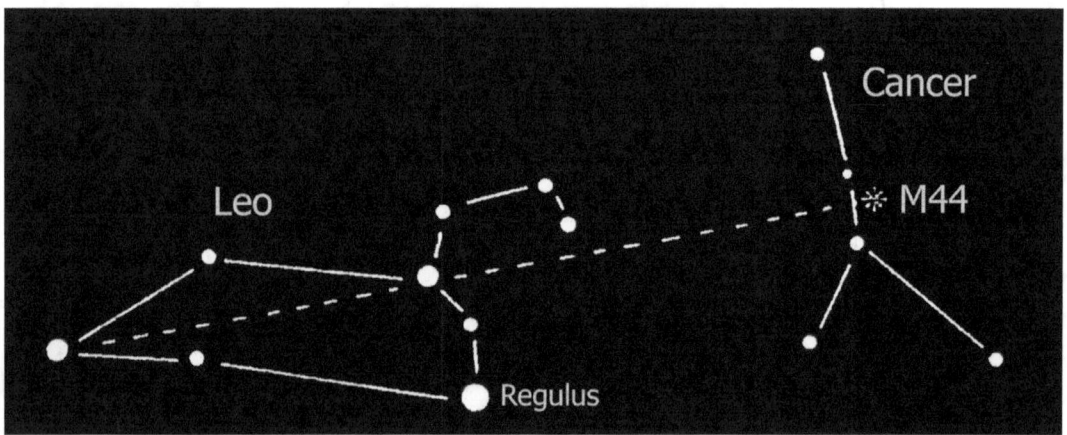

From the Leo Constellation, it's easy to arrive at a faint and indistinct luminous spot: it's the open cluster M44, in the heart of the faint Cancer Constellation.

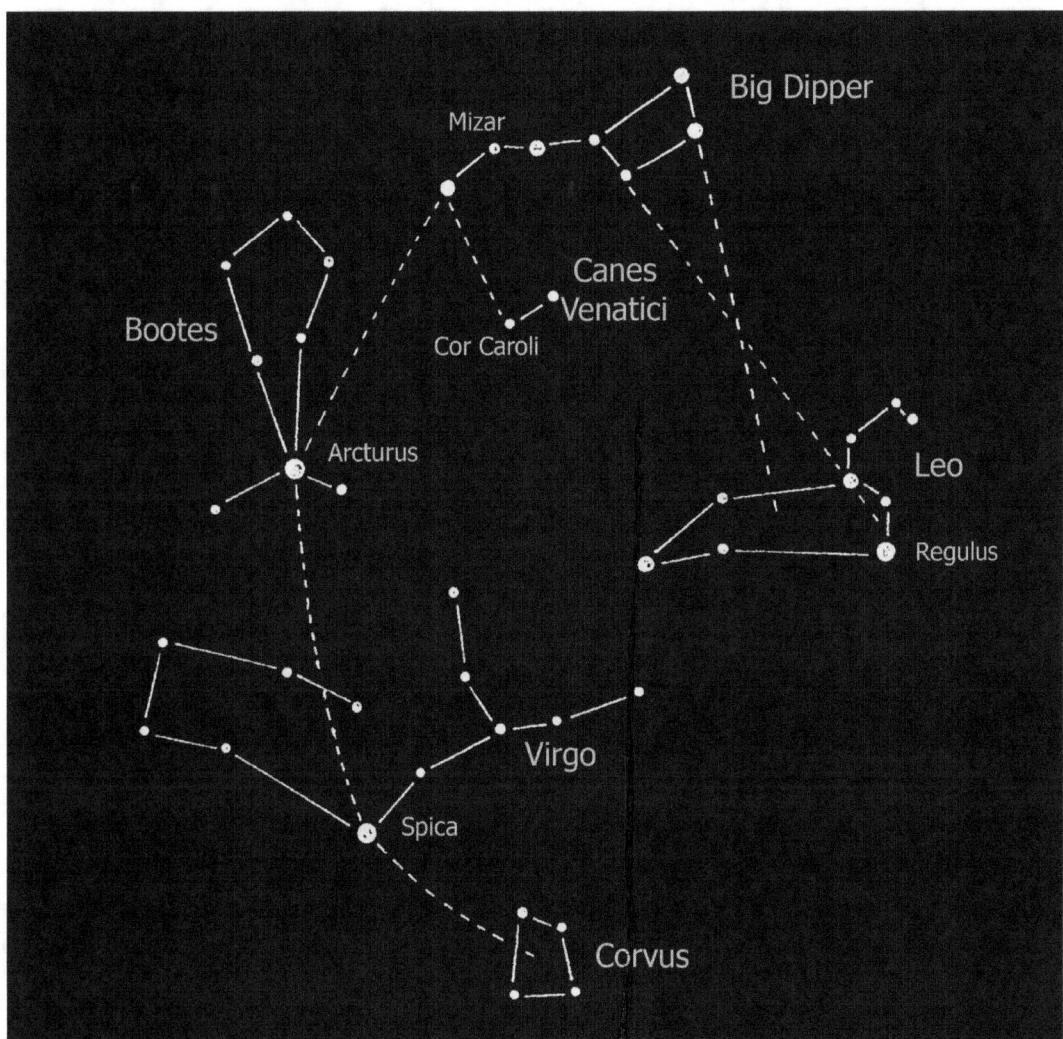

In the spring sky, taking off from the Big Dipper, way up overhead, we can easily track down all of the principal constellations. Extending the arc formed by the 3 stars in the handle of the Dipper, we reach the brilliant star *Arcturus*, from the Shepherd Constellation (*Bootes*), shaped like a kite. Extending even further, we arrive at a white star, less brilliant than *Arcturus* but evident: *Spica*, from the Virgo Constellation. Below, we can identify the trapezium formed by the Corvus Constellation. Extending the last two stars of the Dipper's handle, the ones that also indicate the position of the Northern Star to us (on the opposite side) we arrive at the obvious Leo Constellation, one of the most beautiful and luminous in the sky. Helping ourselves with complete maps, like the ones furnished on the preceding page, we are able to track down, at this point, all of the constellations of the spring sky.

The Shepherd Constellation (*Bootes*) dominates this image with the brilliant Arcturus below. The figure, shaped like a kite, is developed at the top. To the left, the Corona Borealis can be seen. If you don't recognize the figures, try to help yourself using the map of the sky on the next page.

The summer sky

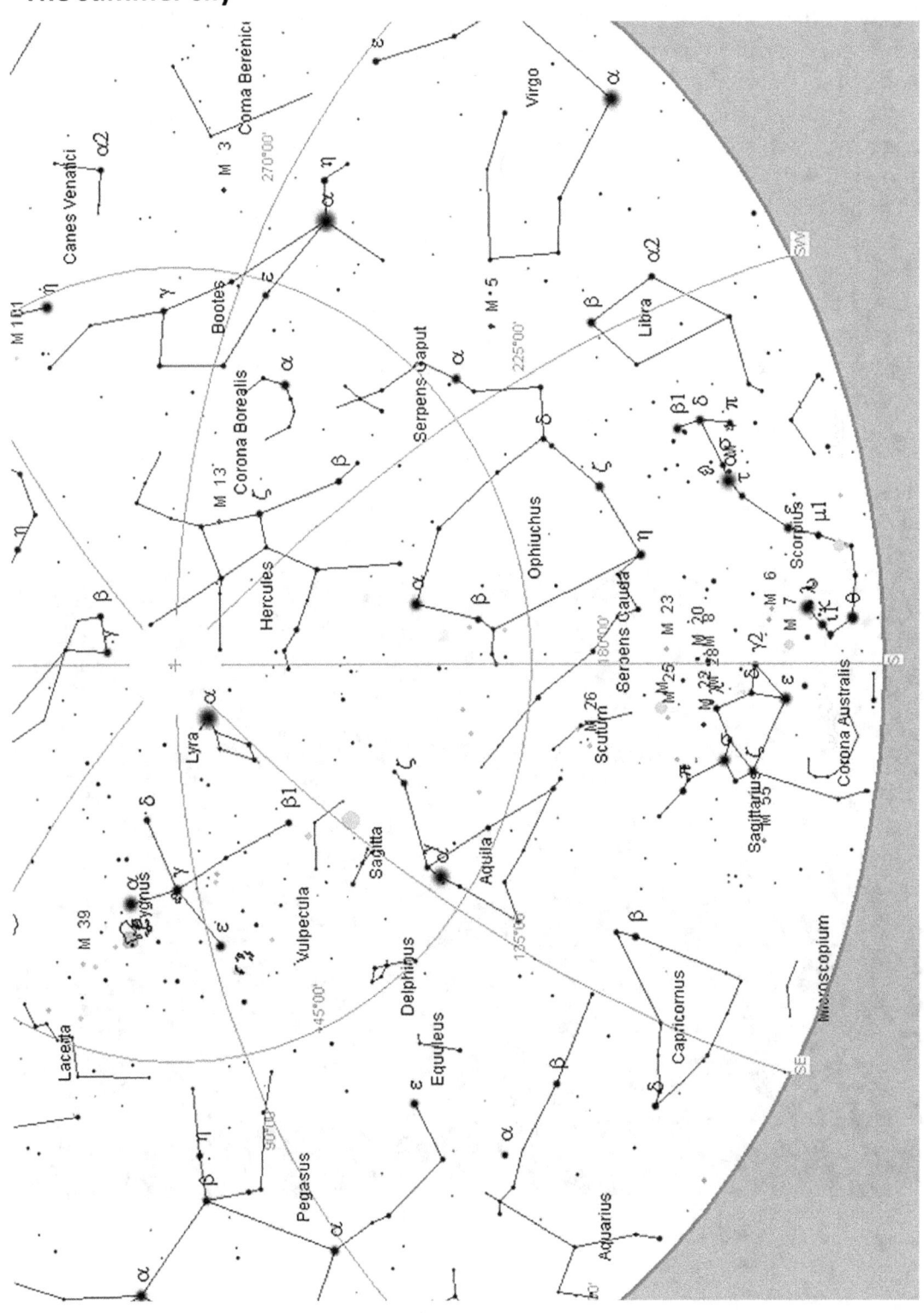

The sky of the hot summer nights is the most beautiful of the year.

Toward the middle of July, when the light from the sunset finally disappears after 10:00 pm, we can enjoy the display of the Milky Way that plows through the sky from north to south, from the Cygnus Constellation to those of Sagittarius and Scorpius.

Due south, not too high on the horizon, you'll see a rather bright, unmistakably orange star: this is *Antares*, the heart of the Scorpius Constellation, among the easiest and most suggestive figures to be recognized in the sky.

To the left, that is, to the east, we find the center of our Galaxy and the Sagittarius Constellation. This zone has a wealth of star clusters and nebulae, some of which, like the Laguna nebula (M8), are visible even to the naked eye.

Overhead, you'll notice a brilliant white star that will accompany you for several months: this is *Vega*, the most luminous star of the summer evenings, which is part of the Lyra Constellation. *Vega* is a perfectly white-colored star with its magnitude equal to 0. Not far off, in the middle of the Milky Way, there is another brilliant star, although less visible than *Vega*: *Deneb*, from the Cygnus Constellation. Further down *Altair*, from the Aquila Constellation, closes the so-called summer triangle, formed by these three brilliant stars which are easily observed even in skies that aren't perfectly dark.

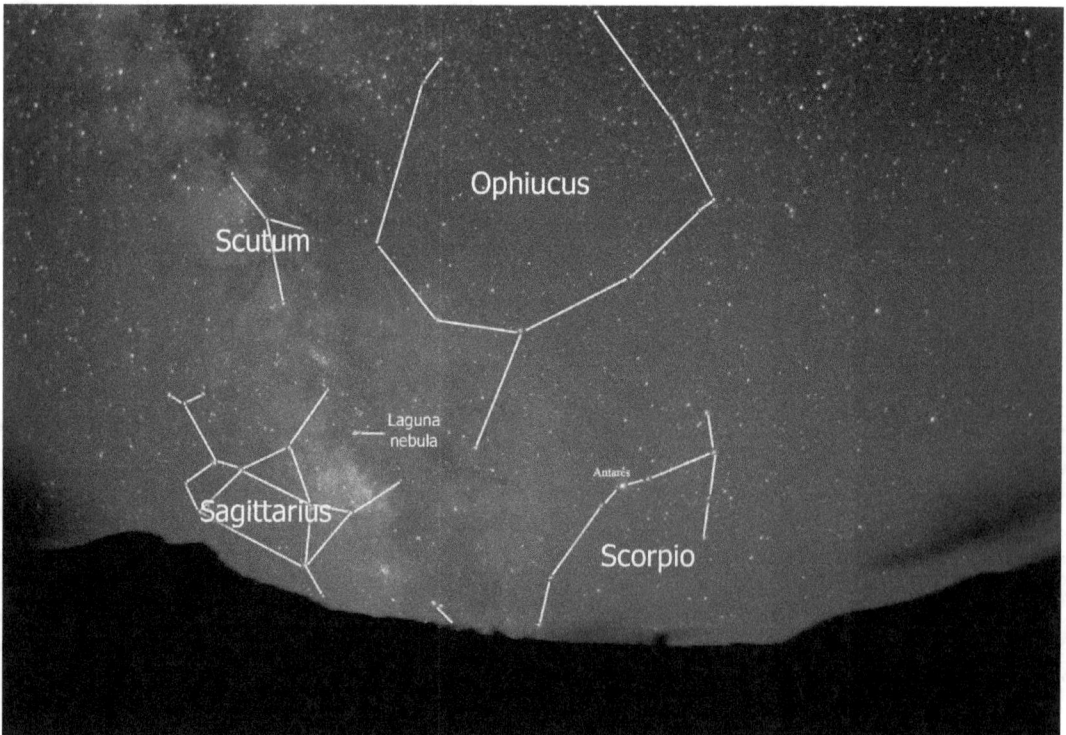

View of the summer Milky Way from a dark sky, looking southward at 11:00 pm on July 15. This is more or less the view as seen by the naked eye.

The summer triangle is the unmistakable asterism of the summer nights. Formed by the stars *Vega* (Lyra), *Deneb* (Cygnus) and *Altair* (Aquila), it is the center of the Milky Way disc and allows us to identify numerous constellations. Paradoxically, it's fairly hard to identify in a truly dark night because of the large amount of dim stars, structures and nebulae that are present in this heavenly zone.

The autumn sky

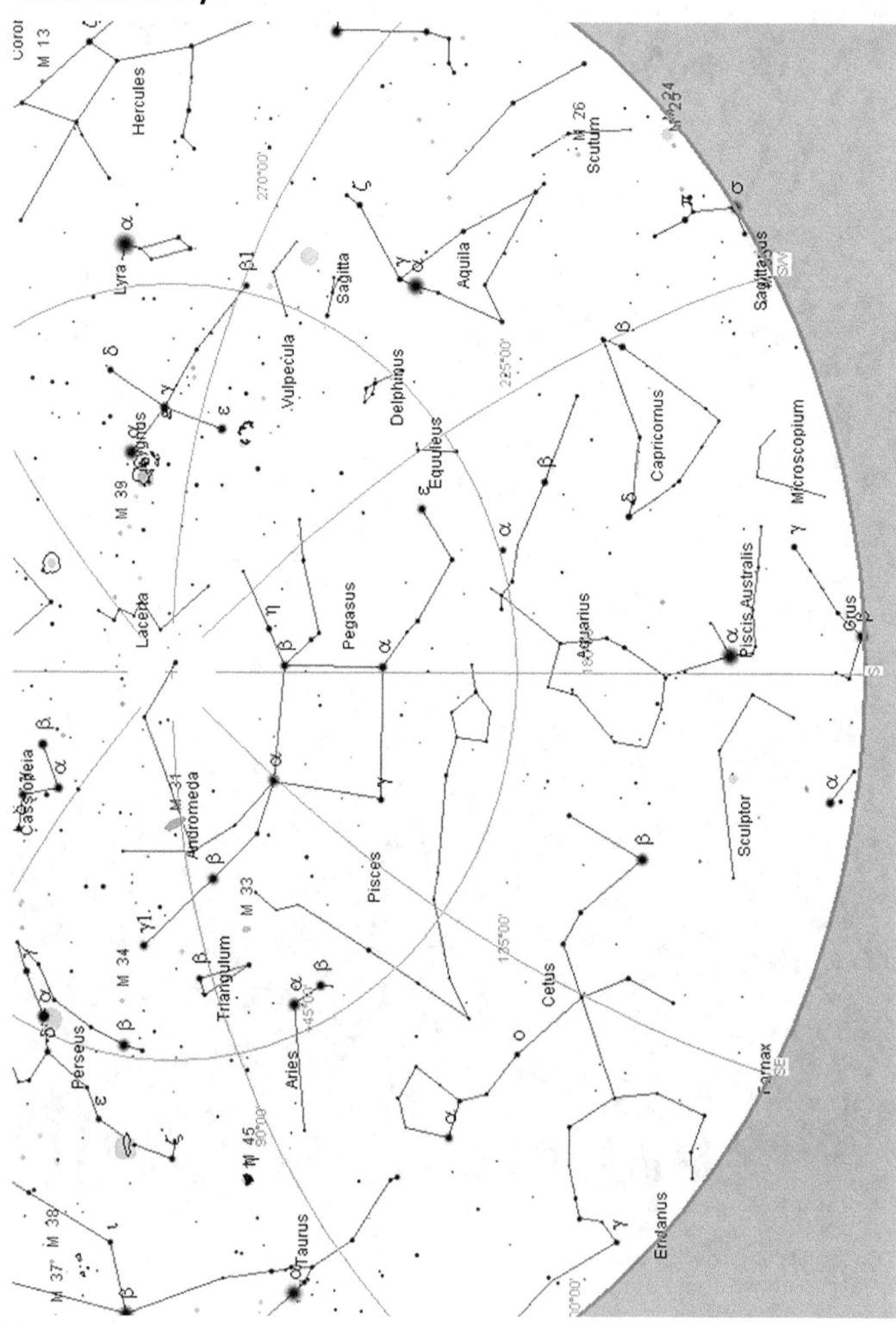

The autumn nights mark the transition between the summer sky and the winter sky.

During the months of October and November, around 10:00 pm, to the west and southwest, you'll find the summer constellations that are getting ready to set.

To the east Taurus, Gemini and Auriga signal winter's approach. After 11:00 pm Orion will also begin to rise, followed, about half an hour later, by *Sirius*.

Directly overhead, we can recognize the Andromeda Constellation in which we can observe, without a telescope, the great Andromeda galaxy, the extragalactic object closest to us, despite its distance of 2.3 million light years.

The light that we observe left the more than 200 billion stars that form this galaxy when there were still no traces of civilization and the first Australopithecines (man's ancestors) were crossing the African steppes.

Attached to the Andromeda Constellation we find the large square of the Pegasus Constellation which, to tell the truth, doesn't much resemble the famous flying horse. Interestingly, the two constellations share a star; it is the *alpha* from Andromeda, the brightest one.

Toward the north, the Big Dipper is found on the lowest point on the horizon. Since it is a circumpolar constellation, it never sets at our latitudes.

Directly opposite the North Star, and, therefore, very high in the sky, we can find the beautiful and unmistakable *W* of Cassiopeia. Nearby, you'll find Perseus, another stupendous constellation; both are immersed in the winter Milky Way – less spectacular than in summer but still interesting – which will, in the course of the following weeks, relentlessly take the place of faraway galaxies that the autumn skies give us.

The *Auriga* Constellation, characterized by the bright star *Capella*, is easy to identify in the autumn sky. Located in the center of the Milky Way, the constellation has a wealth of bright stars, nebulae and open clusters, some of which are visible even to the naked eye. Photo courtesy of Davide Trezzi.

The Taurus Constellation is dominated by the bright *Aldebaran*. Above, the Pleiades resemble a tiny dipper.

The Pleiades as they appear to the naked eye under a very dark sky. Their form brings to mind a tiny dipper.

The winter Sky

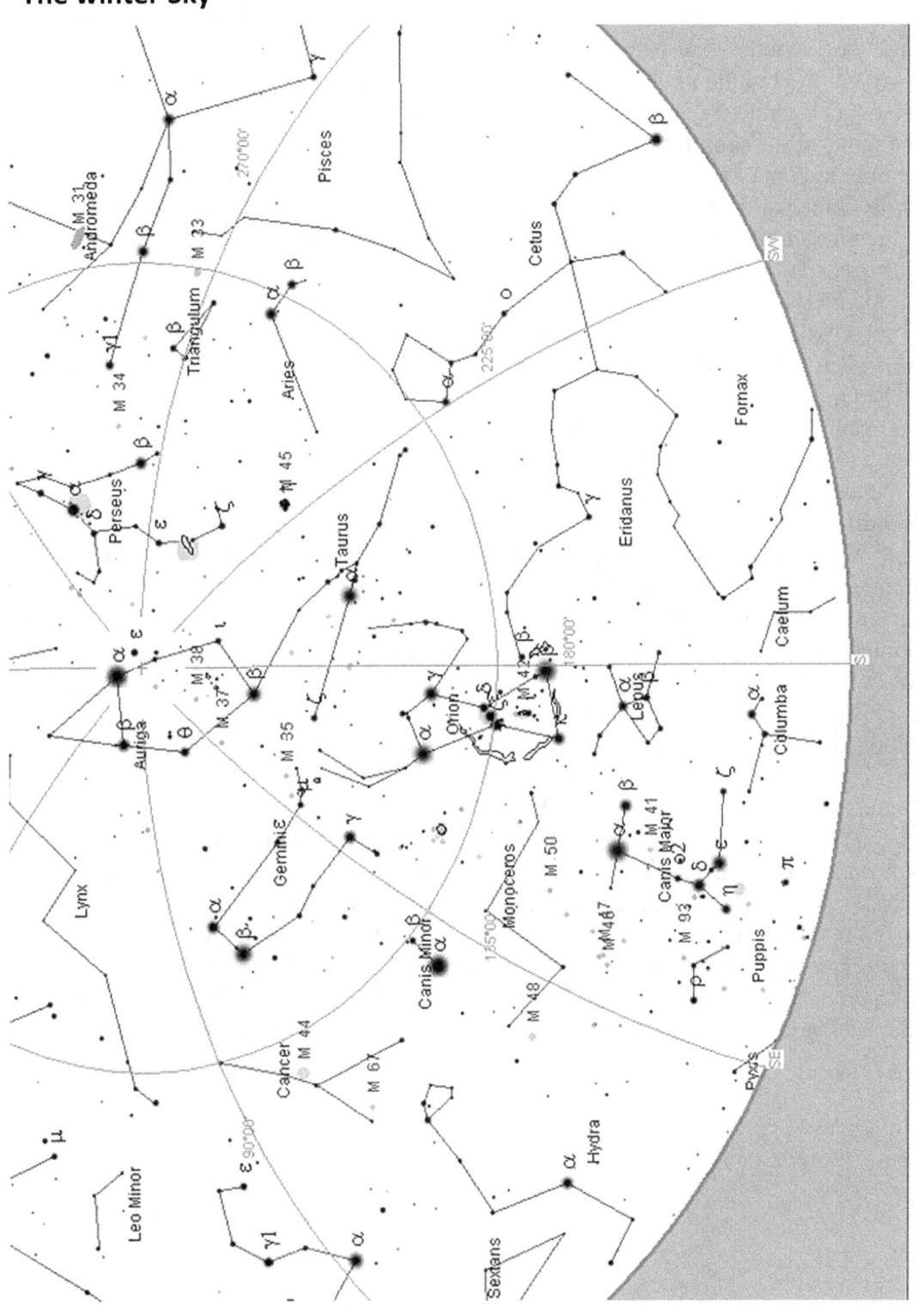

Winter is here, with its burden of cold, but also with crystalline skies thanks to the dry winds. Observations can already begin around 6:00 pm and we can enjoy a long night lasting more than 12 hours.

Around the middle of January, when the Sun has set and the sky has become dark, we notice right away, toward the southeast, a very bright star, the first to appear after the lights of the sunset: this is *Sirius*, the brightest of the entire sky, and yet, it's twice as faint as Jupiter and 14 times fainter than Venus! *Sirius* is, in any case, evident and is the principal component of the Canis Major. This star is one of the closest to the Sun, only 8.6 light years, about 49,700,000,000 miles away: an enormous distance by Earth standards, but not much according to the scale of the Universe.

From *Sirius*, it's easy to point directly to the most beautiful and striking constellation in the whole sky, around twenty degrees to the northwest: Orion, with the three stars of his belt placed unmistakably and excitingly.

The Orion Constellation is one of the most beautiful in this section of winter sky, a real gem that shines above all during the evenings swept by the winds, which clean smog and mists from the sky.

We can find our first nebula in Orion: it's M42, better known as the Great Orion Nebula, identifiable as a slightly blurry star in the heart of Orion's sword, south of the belt, at the center of the constellation. The most interesting star is without a doubt *Betelgeuse* (mag. +0.58, slightly variable), a red giant 100 times larger than the Sun, set on the high left-hand corner of the quadrilateral that identifies the mythological hunter. *Betelgeuse* has reached the final stages of its life, to the extent that astronomers expect, from one moment to the next (from now thru a 50,000 years), a spectacular explosion that, for an entire month, will make it many times brighter than the Full Moon: a farewell in grand style before leaving us forever and rendering the constellation bereft of one of its brightest stars. Who will be the first to see the explosion of Betelgeuse? Considering the Murphy's law, it will be cloudy that day.

On a diagonal from *Betelgeuse* we find *Rigel*, slightly brighter (mag. +0.12), a blue star with a much different age and physical properties; it is the sixth brightest star in the sky. At a similar distance between *Sirius* and *Betelgeuse*, but toward the northeast, we find another fairly bright star, with a magnitude somewhere between *Betelgeuse* and *Rigel*: it's Procyon, from the already mentioned Canis Minor constellation.

Procyon, *Sirius* and *Betelgeuse* identify a triangle in the sky, called the winter triangle, analogous to that of the summer formed by *Deneb*, *Vega* and *Altair*.

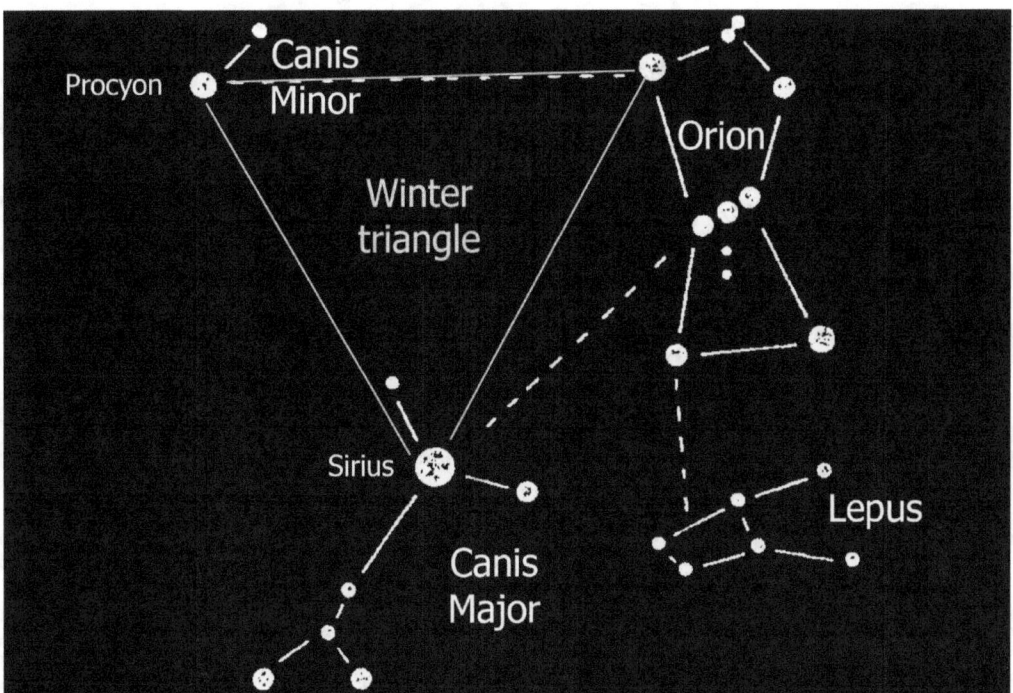

A few movements between the winter stars to identify the brightest constellations of the period. The figures of Orion, Canis Major and Minor are fairly evident and easy to track down without difficulty.

The Perseus Constellation dominates the winter nights.. From its figure, we can track down the *Auriga*, the W of Cassiopeia, Andromeda and, on the opposite side, the unmistakable form of the Pleiades.

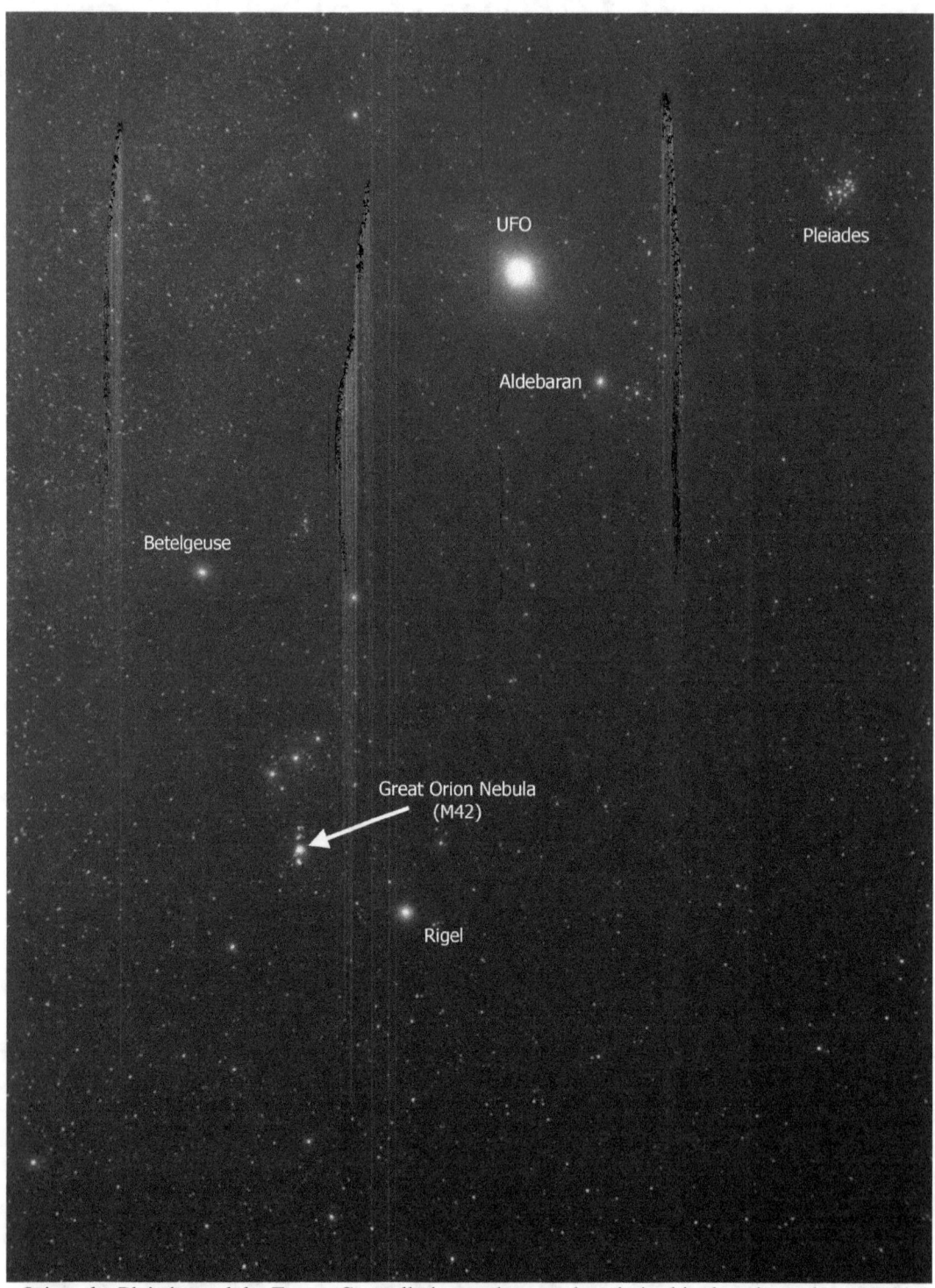

Orion, the Pleiades and the Taurus Constellation as they can be admired in the winter sky, very high on the horizon. The luminous point in the high part is not an UFO but Jupiter which, at the time when I took this photo (November 2012), could be seen in that area.

To the north of Orion we find a few stars characteristic of this season.

The Taurus Constellation, with the red *Aldebaran*, and the small group formed by the Pleiades, also called the seven sisters. A dark sky, and an average eye easily identifies 7 components, while better-trained eyes can easily see 9 and maybe even more.

We find two bright stars to the east of the Taurus Constellation: *Castor and Pollux*, mythological twins that are part of the zodiacal constellation of the Gemini.

Still higher *Capella*, an orange star, dominates the Auriga Constellation, passing almost exactly over its *zenith*. In this constellation, in the midst of the winter Milky Way, we can observe numerous open clusters, some visible to the naked eye, all of them within easy reach of a pair of binoculars.

Turning your gaze northward, the Big Dipper begins to slowly climb back up in the sky. Perseus is almost right overhead, not far from Auriga, with the double cluster clearly visible to the naked eye.

Cassiopeia, on the hand, is slowly descending toward the horizon, which it will never reach in any case, as we well know: it's a circumpolar constellation.

The first observation

Well, the time has come to go back outside to observe, but this time we won't be satisfied with just finding our bearings and finding the North Star. This evening's objectives are the constellations! What do we need for this first official outing? Certainly the accessories from the previous evening: compass, red light and maps. Those mentioned in the preceding pages are good for finding the mentioned constellations, but if we want to be sure we can make them ourselves with the programs I have already mentioned. In particular, SkyChart, easily downloadable from this link: http://www.ap-i.net/skychart/start is a lightweight program that operates on any computer, easy to set up even for those who don't know anything about informatics. It's best to print more maps than necessary rather than arrive in the field unprepared, because then it will be hard to find sheets of paper in the middle of the country!

It would also be great to be armed with an astrolabe, an object made of construction paper which allows you to view, without going to the computer screen, the sky as visible at the chosen time and period of the year. You can find them in a bookstore for a low price or we can build our own with a little cardboard and by following the instructions in internet. This is what all the books I began reading suggested, when I understood that, before using a telescope, I had to know a little about the sky. But those were other times; another millennium, in fact.

Now there's an alternative and it's probably better known than those strange, heavy things called books. All the smartphones based on systems like Android, Windows, Apple, Nokia and Samsung allow us to download, often for free, very useful applications that simulate the sky. A few names for directing research might be Sky Map and Star Odyssey for Android, Star Walk for Apple (IPhone, IPod and IPad).

These programs not only display the sky but, depending on where we point the phone, they are also able to recognize the constellations in front of us and therefore represent a huge help in discovering our first celestial figure. As a long-time amateur astronomer who did everything with paper and astrolabe, I suggest that you use tablet and smartphone screens under the starry sky with caution because their light, even if set at a minimum, eliminates our adaptation to the dark and causes a nice bright square to appear for at least a couple of minutes after we have used them.

After completing the list of things we need, let's move on to the aspect that regards comfort. In fact, this time we'll have to stay under the sky for a little longer and if we want to recognize winter constellations like Orion, the cold will be our number one enemy.

Staying still under a calm sky exponentially increases the sensation of cold, which won't take long to become discomfort and make us run for the warmth of our own beds.

One of the biggest mistakes made when just starting out is underestimating the environment in which observations are done. I can usually resist in just my shirt-sleeves at temperatures up to 59°F, but for observations from September to May, I wear a heavy (but comfortable) snowsuit and winter boots, even if the temperatures go over 68°F during the day.

In the winter, with the thermometer that can easily go below freezing, staying well-covered becomes a necessity if we don't want to get sick and lose our passion for the sky. It's better to be overdressed and then remove something during the evening rather than regret that comforter you left at home.

I still remember perfectly a night in the winter of 2003-2004 when the temperature went down to 15°F and a layer of frost settled on my snowsuit and on my hair, making me wish for a nice warm woolen cap that would have saved from unpleasant suffering.

Our feet and hands are the most sensitive parts and should be covered with several layers of clothing. We don't have to worry about esthetics; all things considered, we're going to observe the sky from the middle of a field, at night, far from civilization – not a beauty contest!

A snowsuit isn't enough for the winter nights; we need woolen underclothes for our body and legs and two pairs of wool socks covered by ski boots (those famous moon-boots). Snow gloves and a thermos of hot tea are very useful. The cold, and especially the humidity, should never be underestimated, even in the summer. And, while we're at it, let's take something to eat and a friend or two so we're not alone (and so we won't be as afraid of animals that might possibly approach us during the night).

And now we're ready for our trip toward the dark sky and to put everything we've seen into practice. There aren't any rules to follow, nor advice that I can give. This is an intimate moment that should be lived and enjoyed in a completely personal manner. All that I can do is wish you good observations and to present you with maps that are ready and that we can photocopy or keep with us, depending on when we decide to observe.

Some information on how to use the following map correctly. The pages on the left (the even pages) show the sky as it would be seen if we were turned northward, up to about the zenith (the vertical point above our head), the ones on the left (the odd pages) show the view to be had looking toward the southern horizon, at the same time as the preceding map. The times indicated in the key refer to wintertime; therefore, if we are observing during the summer we have to add an hour to find the aspect of the sky most similar to what is shown. If we want to observe at different times than the ones shown, you just need to understand that the portion of sky observable from any given map will be visible about an hour earlier every two weeks. For example, the first map is valid for January 1 at midnight, but it could also be fine for December 15 at 1:00 am, or March 1, at 8:00 pm (but in this case we might have a problem with sunlight).

Happy observations!

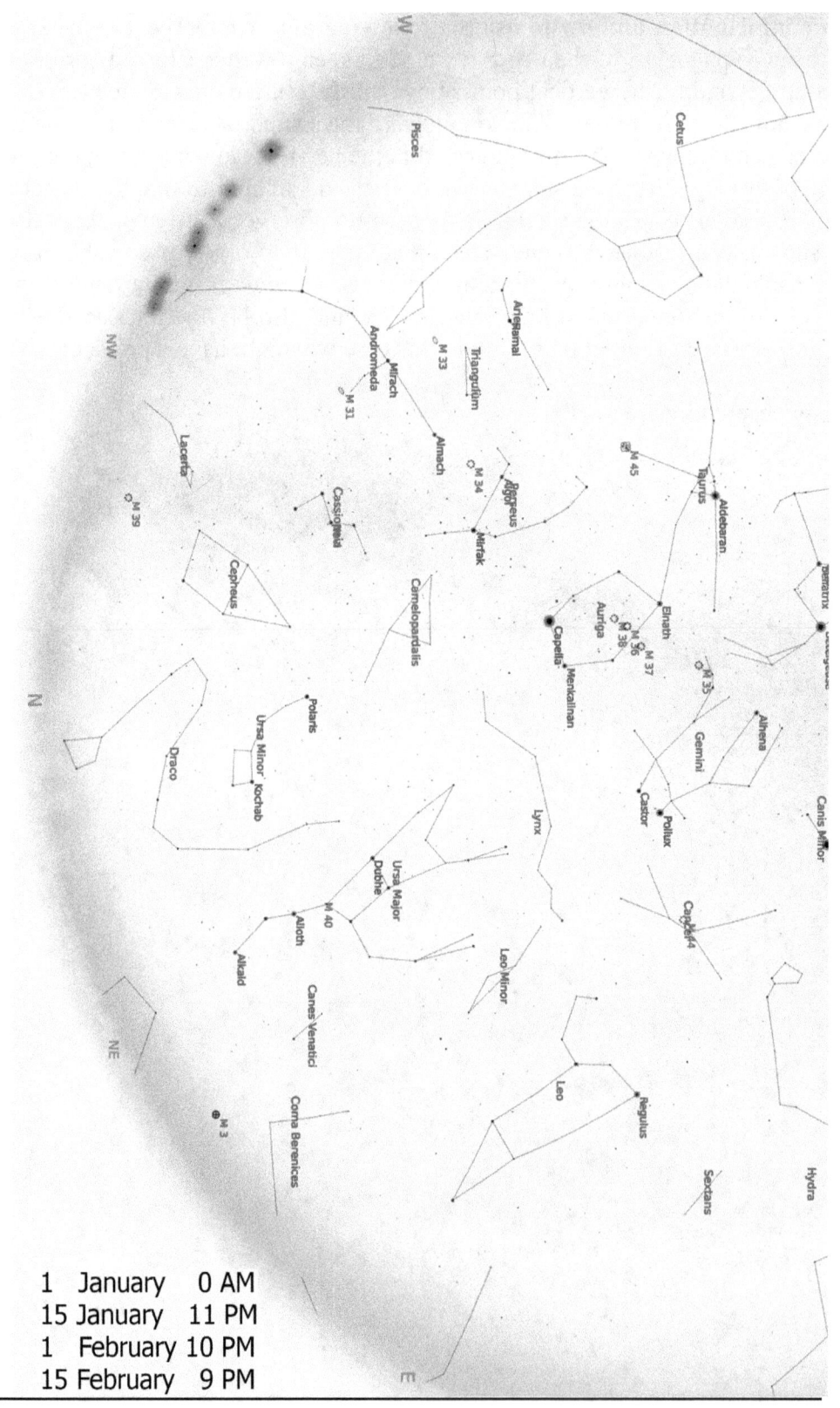

1 January 0 AM
15 January 11 PM
1 February 10 PM
15 February 9 PM

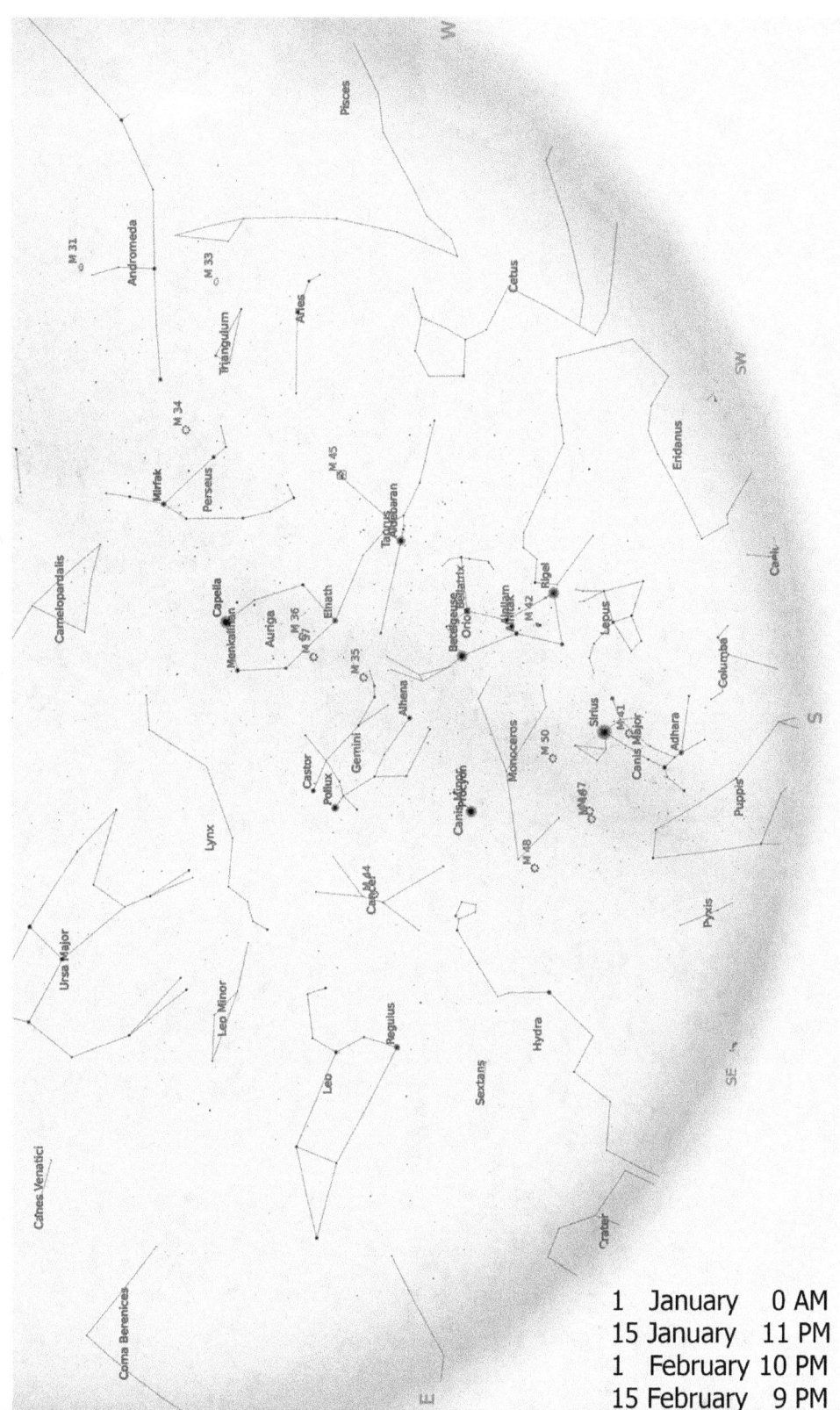

1 January 0 AM
15 January 11 PM
1 February 10 PM
15 February 9 PM

The sky seen with the naked eye — Oh Wow! I Saw Saturn!

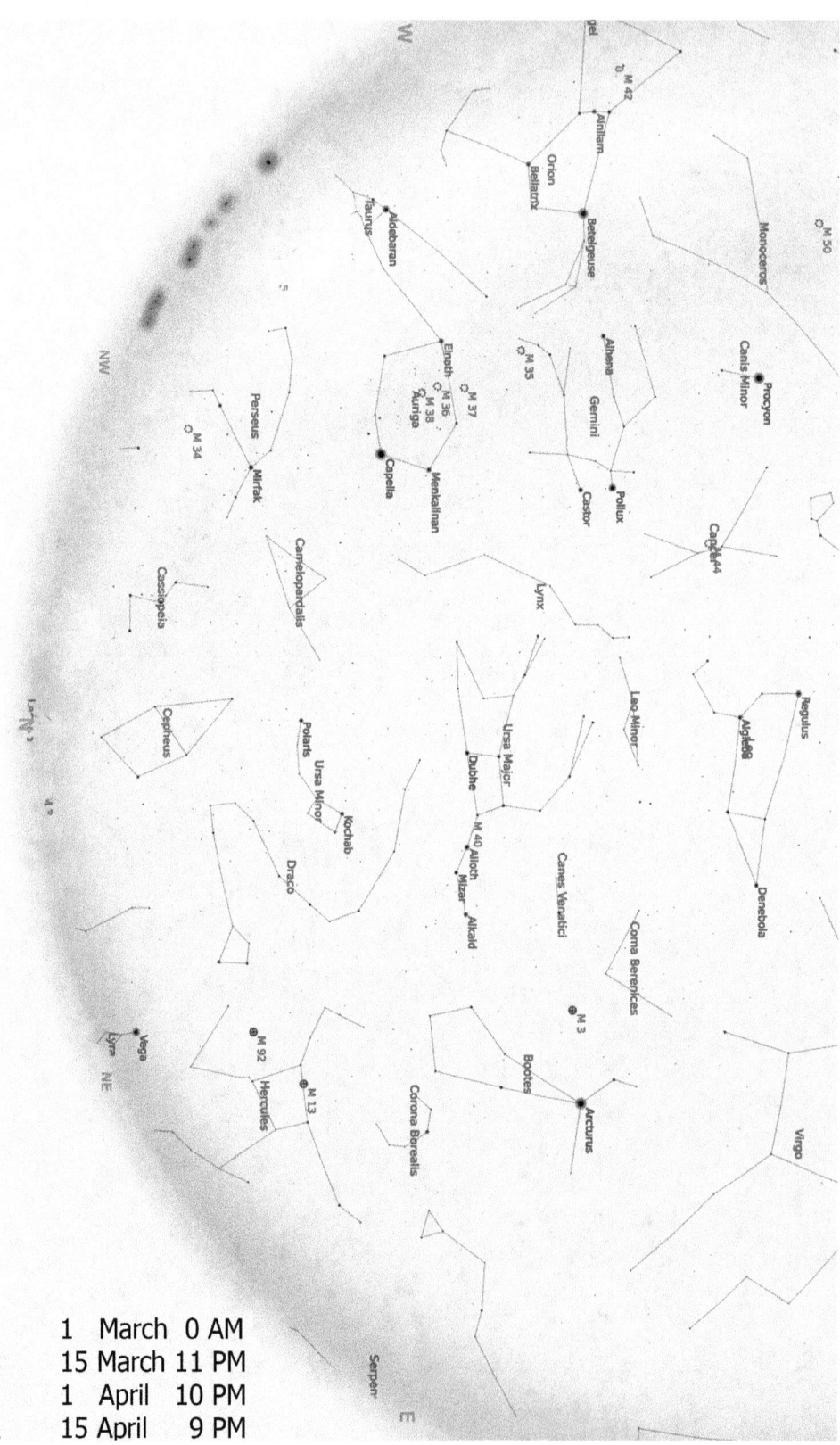

1 March 0 AM
15 March 11 PM
1 April 10 PM
15 April 9 PM

The sky seen with the naked eye — Oh Wow! I Saw Saturn!

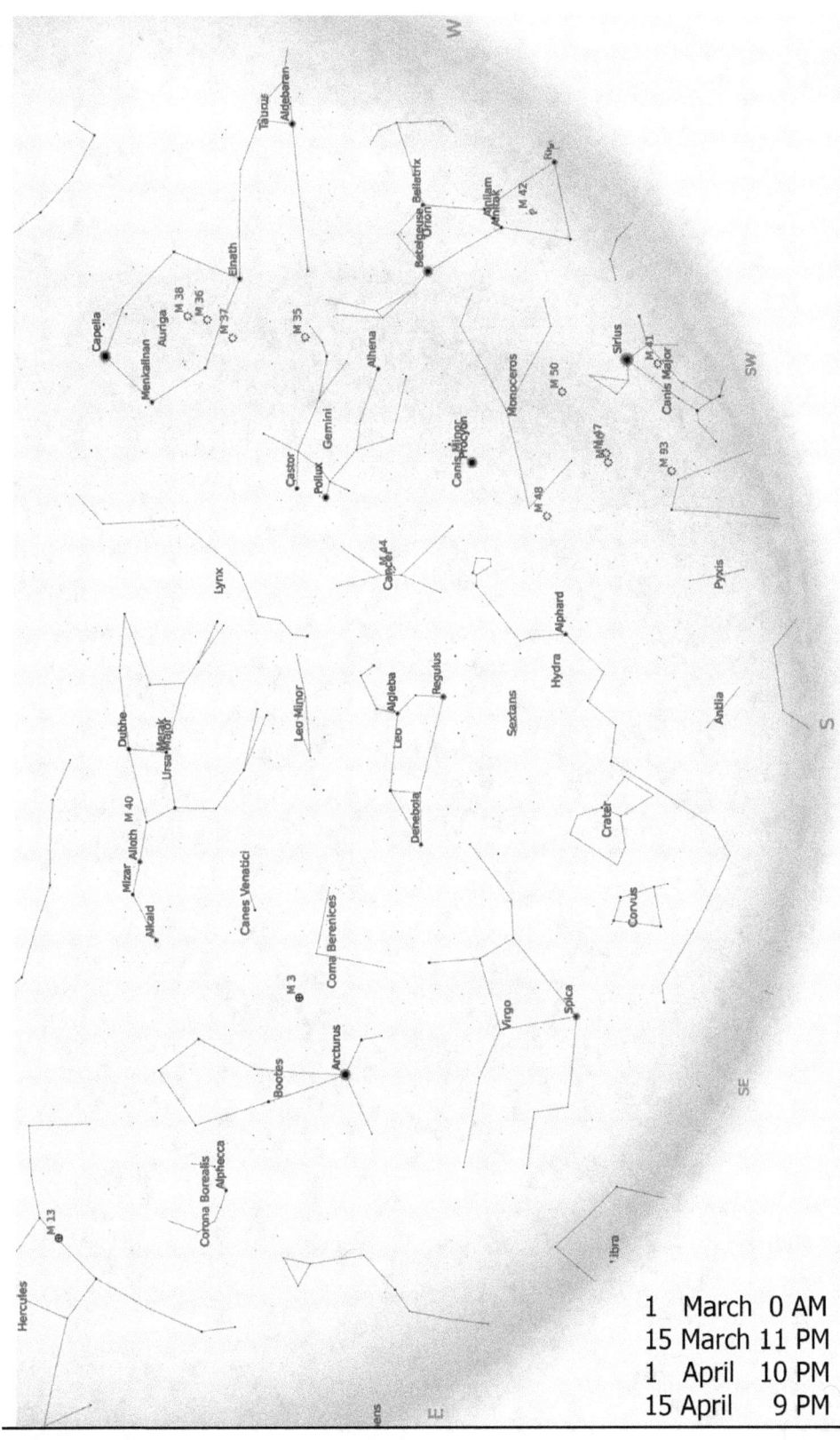

1 March   0 AM
15 March 11 PM
1 April 10 PM
15 April   9 PM

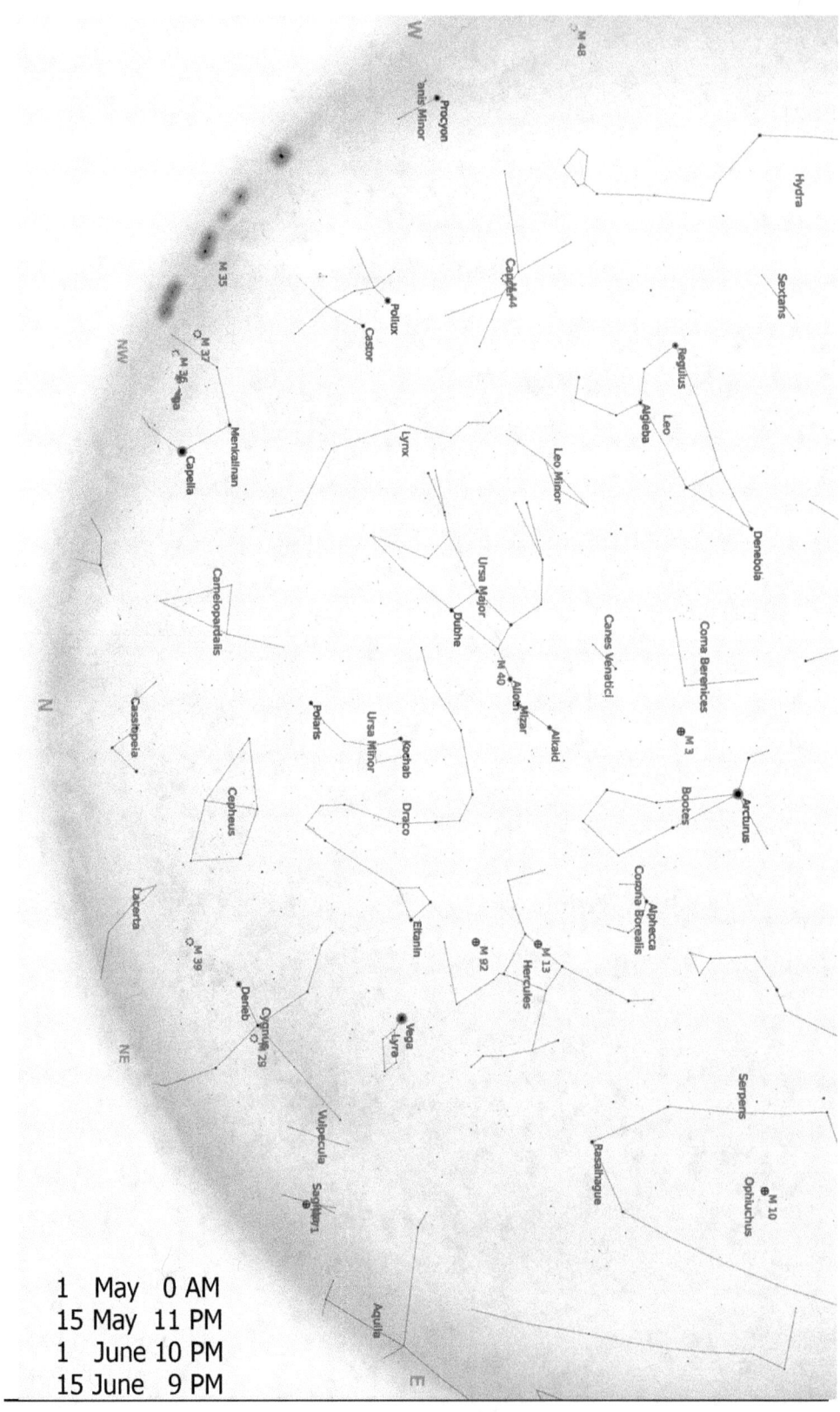

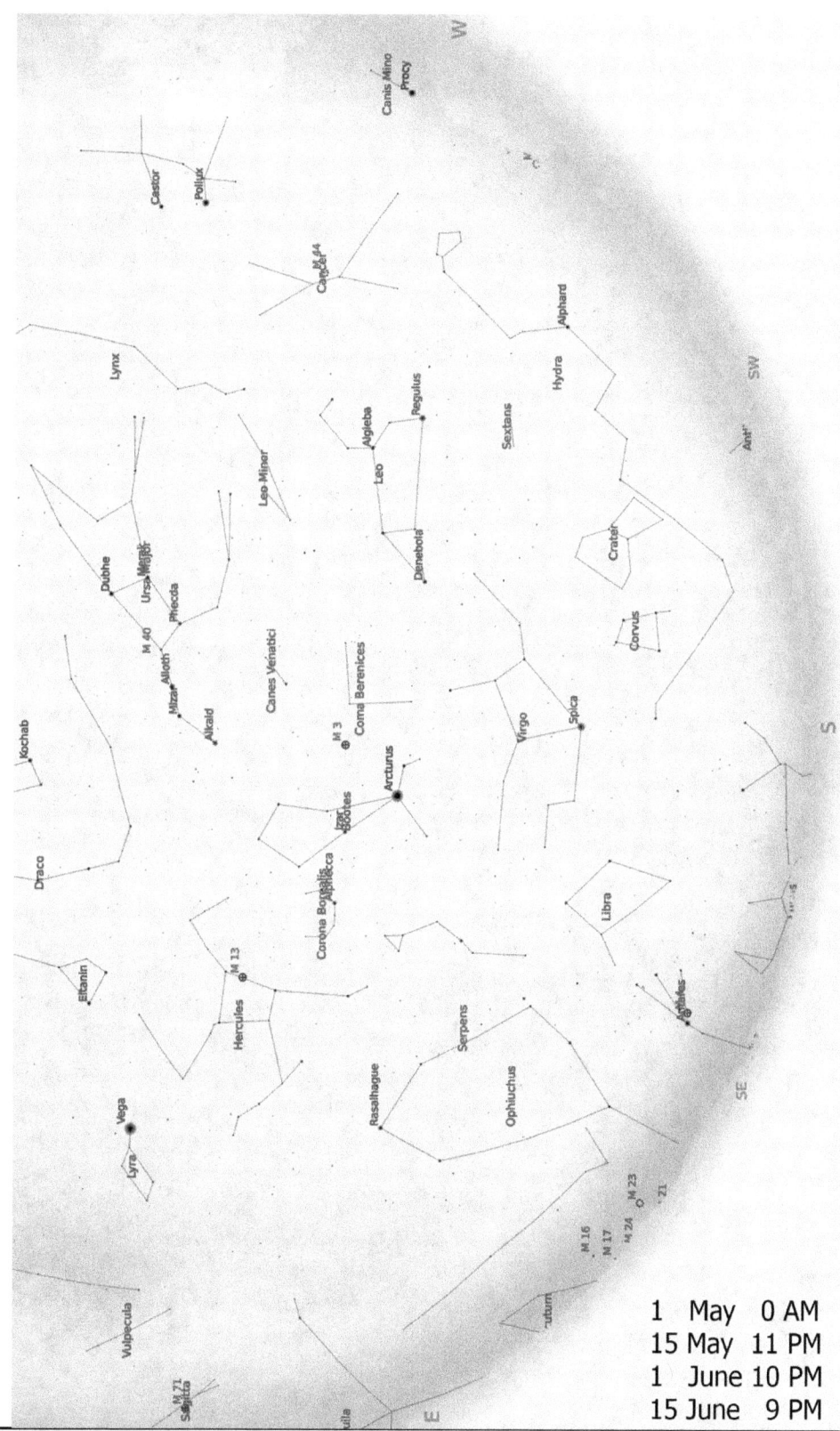

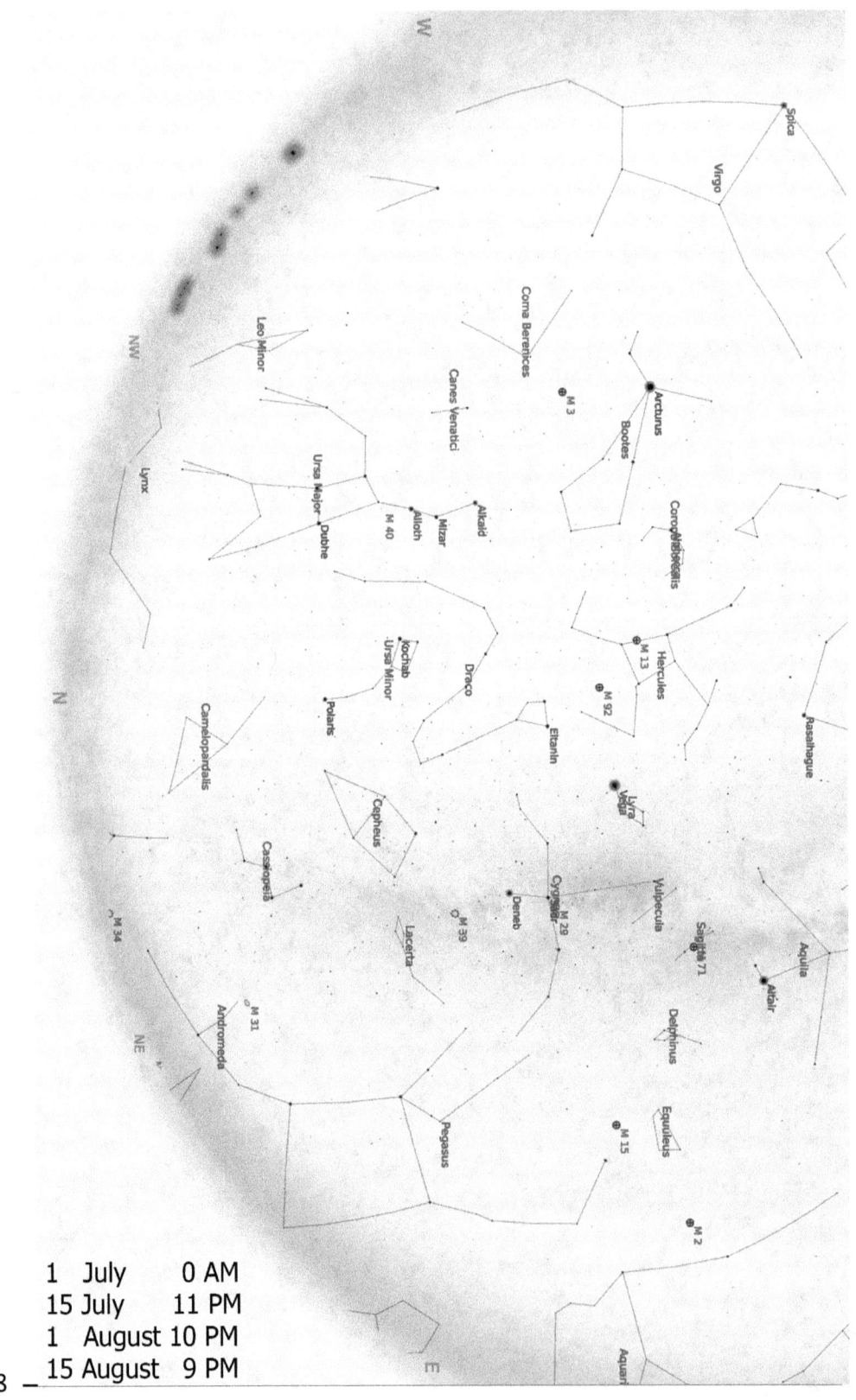

1 July   0 AM
15 July   11 PM
1 August 10 PM
15 August 9 PM

The sky seen with the naked eye

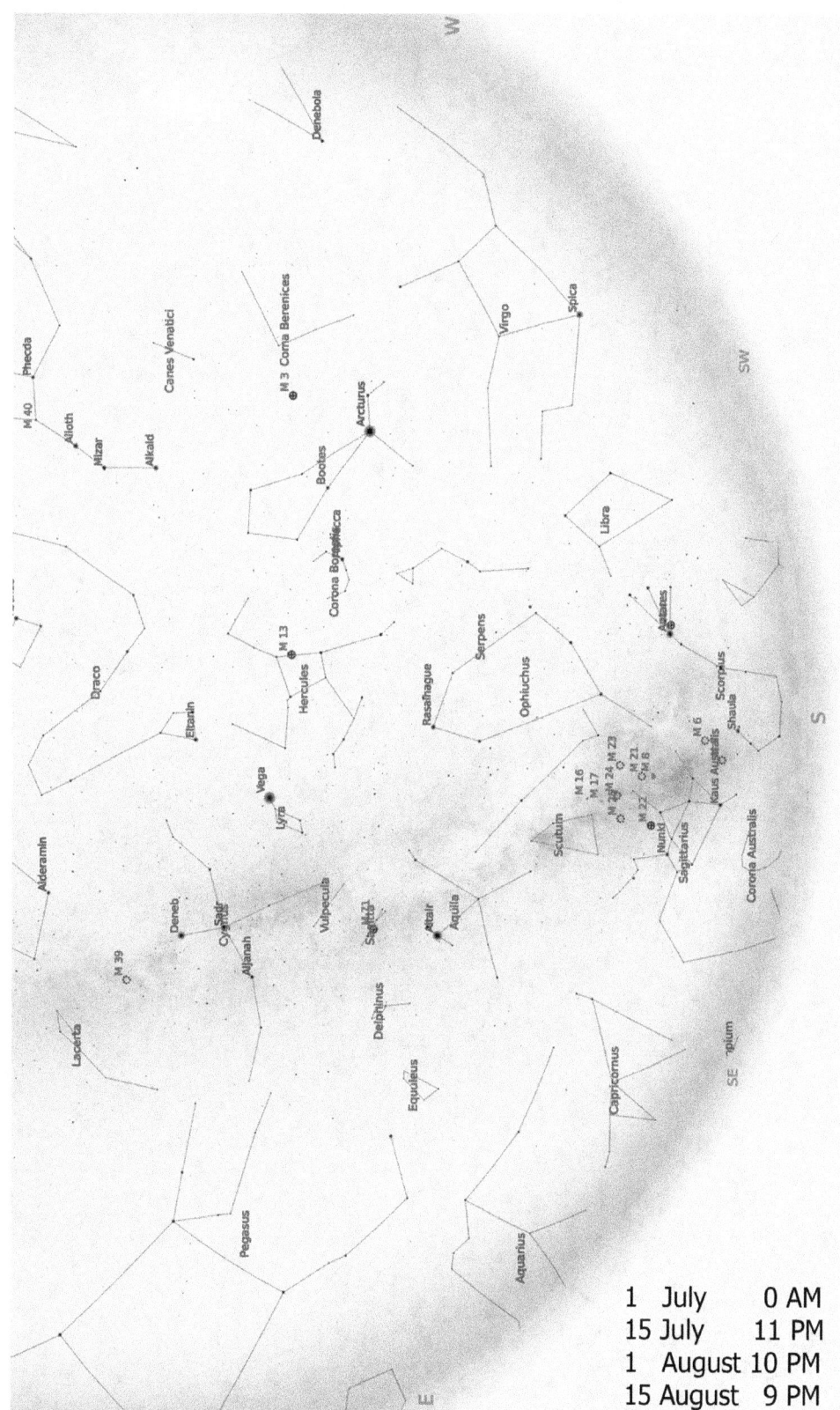

1 July	0 AM
15 July	11 PM
1 August	10 PM
15 August	9 PM

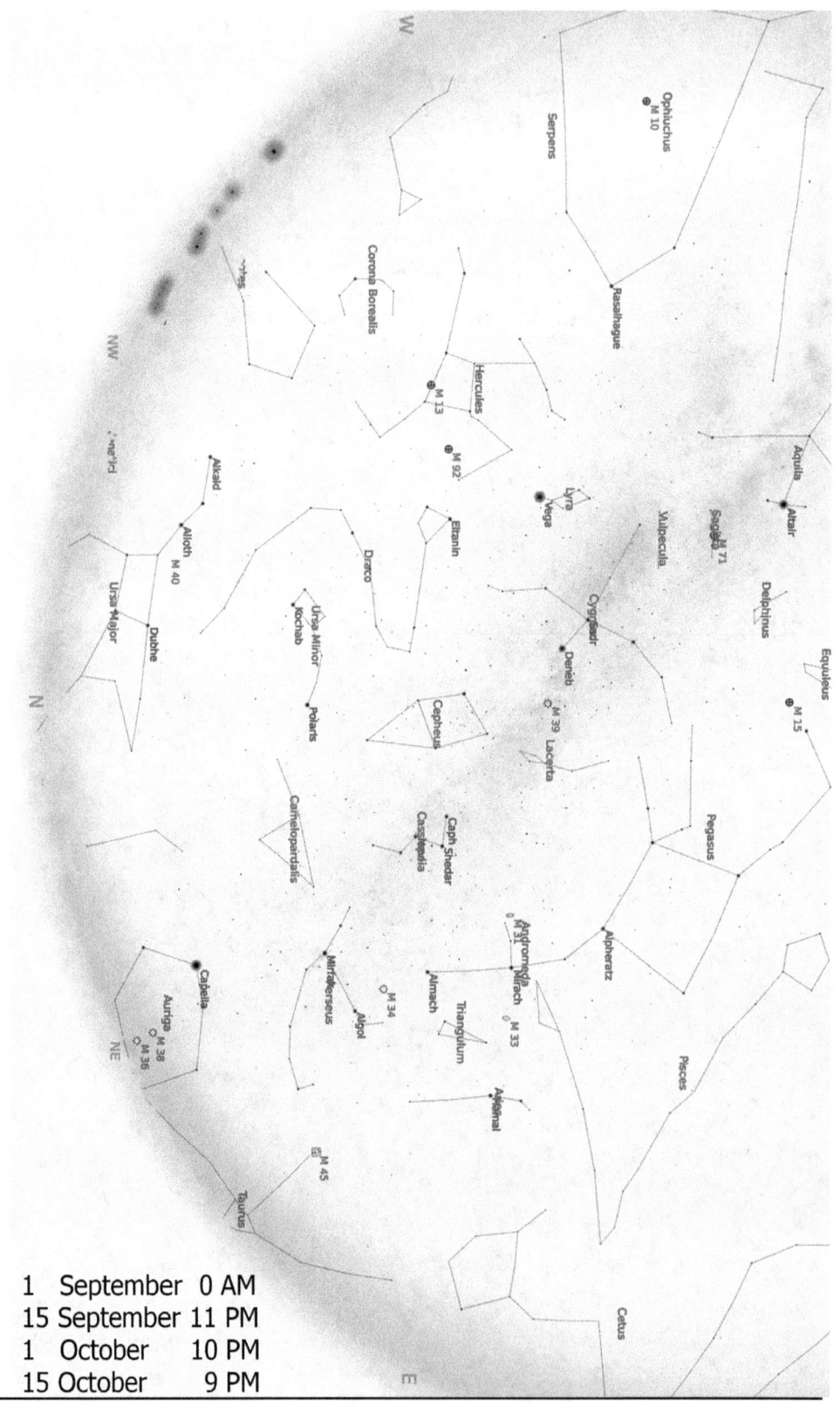

1   September   0 AM
15 September 11 PM
1   October   10 PM
15 October   9 PM

The sky seen with the naked eye — Oh Wow! I Saw Saturn!

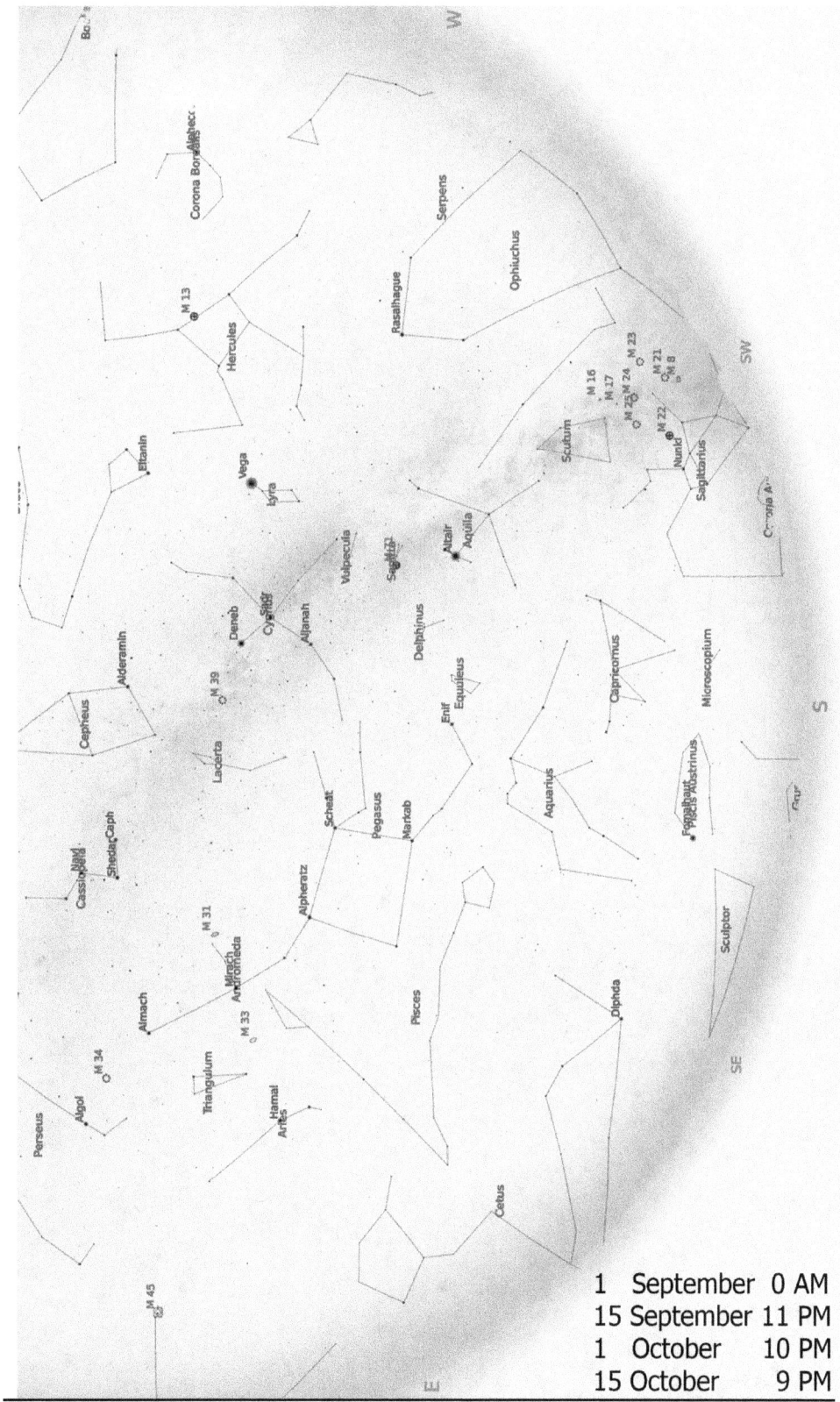

1 September 0 AM
15 September 11 PM
1 October   10 PM
15 October   9 PM

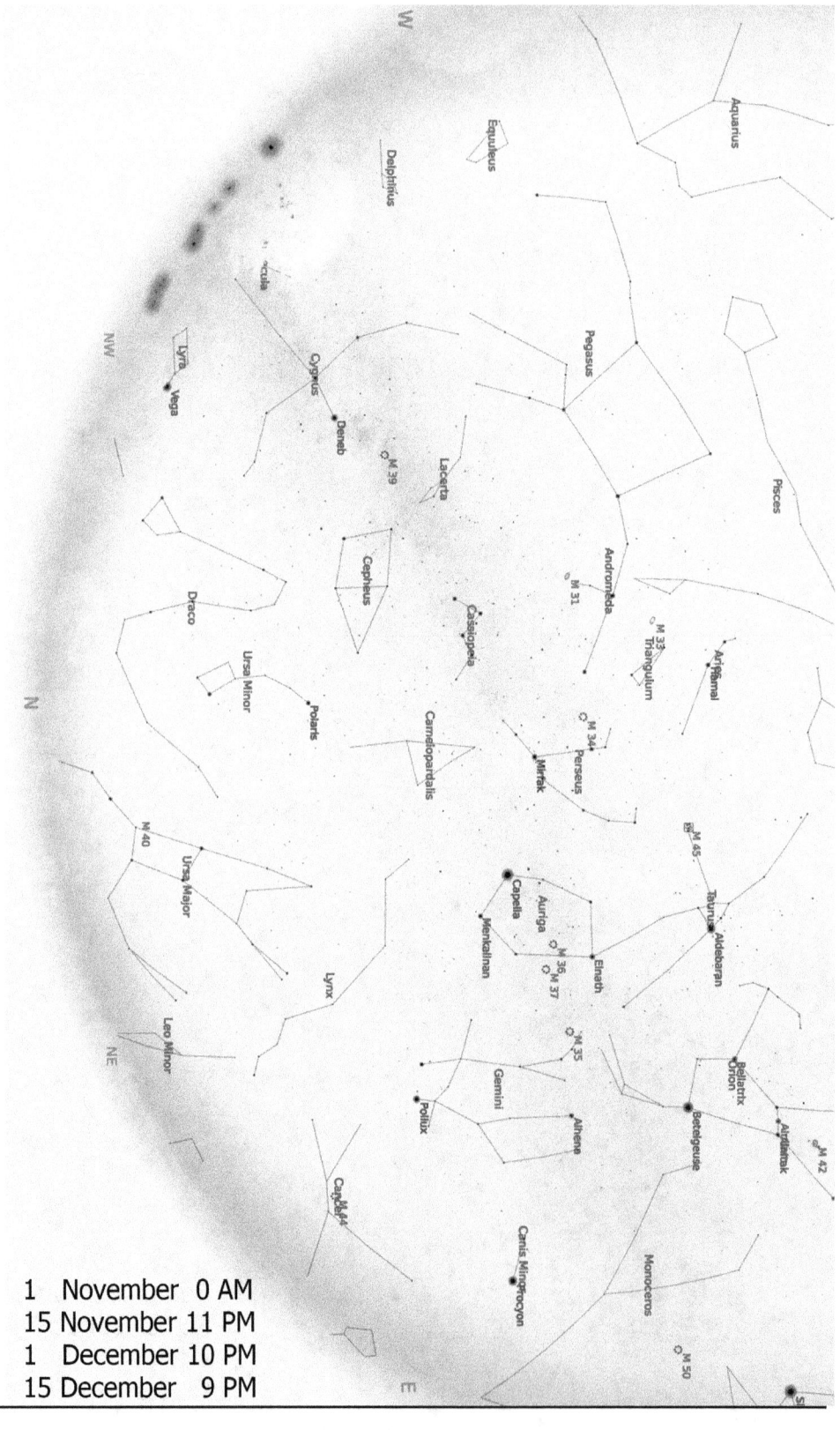

1   November 0 AM
15 November 11 PM
1   December 10 PM
15 December 9 PM

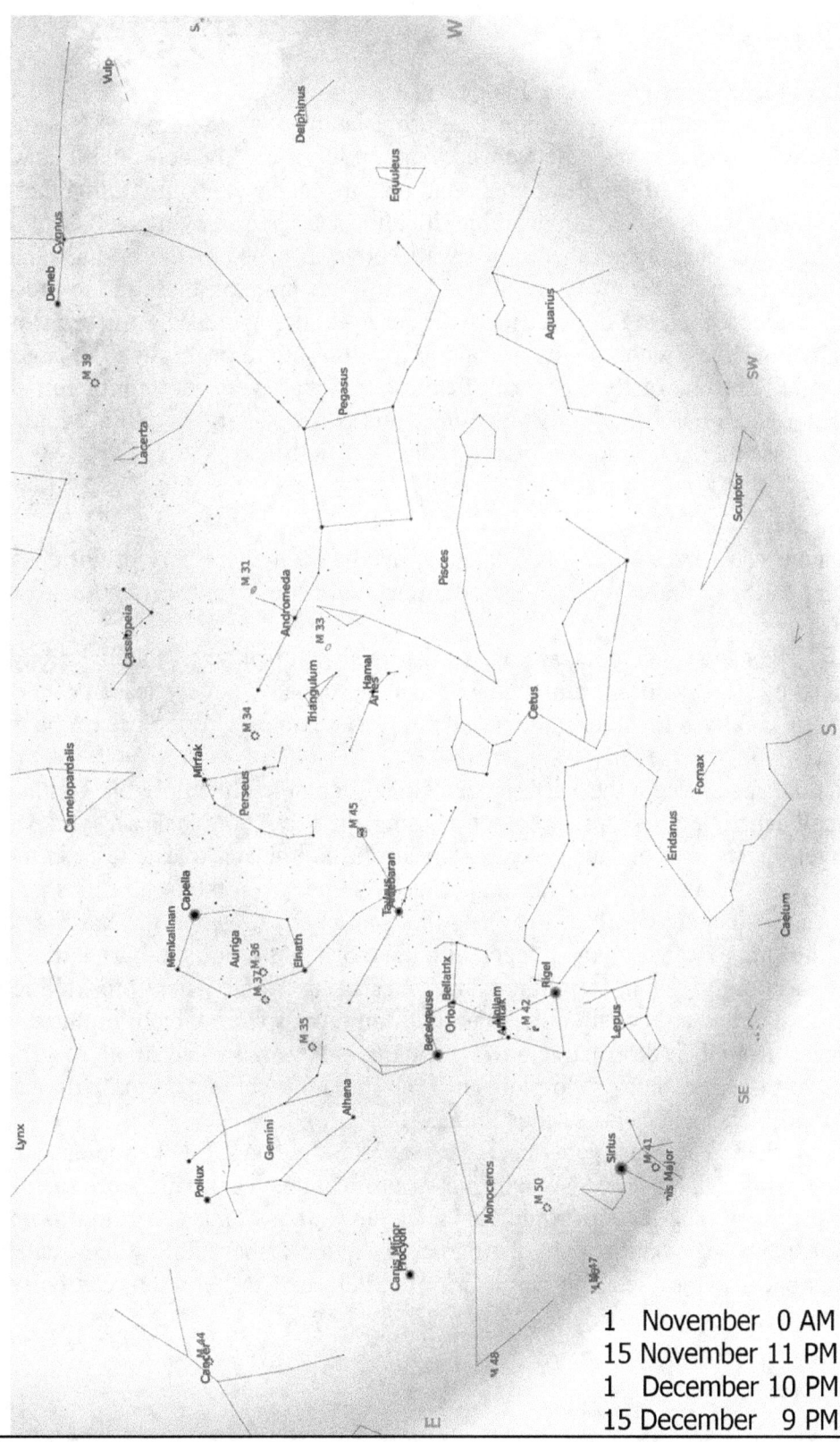

1 November 0 AM
15 November 11 PM
1 December 10 PM
15 December 9 PM

Not just stars

How did the evening go? Good, I hope.

If the sky was dark, the Moon didn't disturb us and we watched for at least an hour, we'll have noticed, maybe while we were hunting for one of the constellations we had already seen, other curious phenomena that perhaps made us shout out that there was a UFO. Don't worry, we're not dealing with alien spaceships spying on us from outer space, but with phenomena that aren't well-known because of the bright lights that blind us and keep us far away from the sky. I didn't mention them in the preceding pages: I wanted them to be identified with our eyes. It's nice to discover unexpected things by ourselves, without anyone having to tell us aloud or through the pages of a book, right? This is exactly the spirit of exploration that every astronomy buff should have in order to enthusiastically undertake this extraordinary adventure among the stars; because there are many surprises to discover in the sky!

Falling Stars

Of course you've seen a few falling stars, maybe even some very bright ones. The latter are called fireballs and can even illuminate the countryside around us for a few seconds.

The falling stars, or meteors, are one of the loveliest displays to watch, strictly with the naked eye (how can you aim at one with a telescope?) in every season of the year. They aren't really stars that fall, but something that happens very close to the earth's surface, on the average only 50 miles up.

In fact the space where the Earth is located isn't completely empty but contains gas and small debris, left perhaps by comets passing close by. It only takes a solid particle, no bigger than a grain of sand, to cause a beautiful falling star to catch fire in the sky. These debris are, in fact, very fast, superior by some seven thousand miles per hour (much faster than any airplane), to the point that when they enter the Earth's atmosphere, they are reheated by the impact with the molecules of air and burn almost completely, becoming visible as falling stars. Fascinating, right? Therefore, falling stars are space debris that burn up on contact with our atmosphere which, in this manner, protects us from their destructive power. Although they are indeed small, even a grain of sand, if thrown at tens of thousands of miles per hour could pass through our body without problems and cause serious injuries.

Anyway, there's no cause for concern; falling stars are beautiful phenomena and safe to watch. Although they are visible in every season of the year, many more can be seen at certain times. The tears of Saint Lawrence are famous and, according to popular tradition can be seen on the night of August 10. Actually, these falling stars, called the Perseids because they seem to come from the Perseus Constellation, can be seen at their maximum between August 11th and 12th every year.

Another highly anticipated appointment for enthusiasts is the night between November 16th and 17th, when the Leonids, falling stars that seem to come from the Leo Constellation, appear. For the best chance to see them, waiting for the second part of the night, when Leo is high in the sky, is recommended.

A beautiful meteor that I watched and filmed by pure luck, while observing the crystalline sky of the southern hemisphere of Australia. Falling stars, or meteors, are always visible in any period of the year.

Satellites

We won't have seen just these quick flashes of light during our first outing beneath the starry sky. We'll have surely seen some dots that are similar to stars that moved much more slowly. They almost seem like airplanes, but contrary to the latter, they have a set light and don't make noise. Some, incredibly, even change their brightness and then suddenly disappear. It seems to be a big mystery; could they be alien spaceships? Spaceships yes; aliens no. In fact, we've been watching our artificial satellites, at least, the ones that orbit closest to our planet, at a quota of about 250 miles. They don't have beacons but are made visible because at those heights it could be that the sun hasn't set yet, and so they are illuminated. And then, their metallic structure is able to reflect the light from the Sun toward us and that is how they appear out of nowhere.

Their irregular shapes cause the changes in their brightness, even sudden changes like those in the Iridium satellites, automatic spaceships that can suddenly become almost as bright as the full Moon for a few brief instants. The magic ends when they, too, make a complete circle around the earth in just 90 minutes (incredible!) and see the setting Sun. Suddenly, the light from the satellite dims until disappears, right in the middle of the sky.

The natural satellite par excellence, and also the brightest, is the International Space Station (ISS), which becomes visible at regular intervals a little after the Sun sets or before dawn. Its brightness can reach that of the planet Venus, with a value of -4!

If we want to know something more about satellites, these two sites promise (with success) to foresee their passage within one second. Just insert the coordinates of the place you're observing from or else select them from a map: www.heavens-above.com and www.calsky.org.

The crowded sky. The trail of an Iridium satellite on a long exposure photograph.

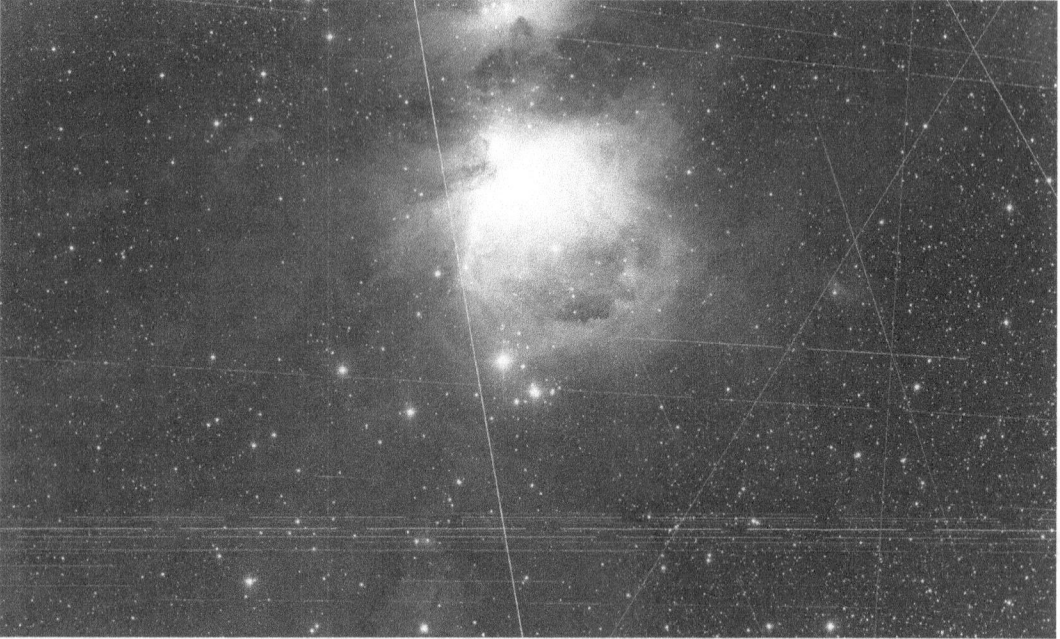

The crowded sky. A huge amount of satellites crosses the Great Orion Nebula in just one hour. Image of Gianni Fardelli, November 2018.

The zodiac constellations

After the pleasant distraction provided by meteors and satellites, the time has come to take another step forward. We'll probably need another night to better understand how the constellations we tried to observe the first evening are made, so it's best to insert another challenge that will be useful for recognizing the planets: let's try to observe the zodiac constellations.

What are they and why are they important?

The zodiac constellations are the designs that the Sun crosses during the apparent path that it completes every year. A lot of difficult words, right? There's really nothing complicated about it. In fact, we've already seen that the Earth makes a complete circle around the Sun in the course of a year, but here on the surface, we see the reflections in the sky. The Earth's orbit around the Sun is manifested, therefore, as a slow movement of our star through the zodiac constellations, along a line – never parallel to our horizon – which astronomers call ecliptic: a complicated name for indicating a simple concept, typical of many scientists!

There are 12 classic zodiac constellations but according to the new boundaries defined by astronomers a few decades ago, there is another constellation that the Sun crosses briefly. The Latin name is Ophiucus, which is also used in English and is translated as the Serpent-Bearer (this is the literal translation).

Why are the zodiac constellations so important? For many people of the past and present, they represent a significance that goes beyond astronomy. The famous signs of the zodiac and the modern horoscopes were born from the assumption that the position of the Sun in a constellation influenced our character and the carrying out of our life.

While this interpretation could be found acceptable a long time ago, when it was thought that the stars were lanterns hung by the gods, now that we have discovered how things really are, it's no longer possible to believe that the stars or the Sun influence our character. And, on the other hand, how could it be possible for our luck to depend on an alignment between the Sun and a group of stars that have nothing in common, which are found at infinitely greater distances and live their existence without even knowing that we are here and that we are observing them?

At this point, the zodiac constellations are important for astronomy buffs only because they are a tremendous point of referral for finding the planets, an endeavor that we will officially try to do a couple of evenings from now.

The zodiac constellations straddle the ecliptic which, in the maps of the sky that we used in our first hunt for the constellations, is marked as a continuous line that crosses the sky at different heights of the horizon. In our latitudes, this line is never circumpolar. This means that none of the zodiacal constellations will be near the North Star, the Big Dipper or Cassiopeia. They will all be arranged toward the southern horizon, rising among the 23° for the summer constellations and almost 70° for the winter ones.

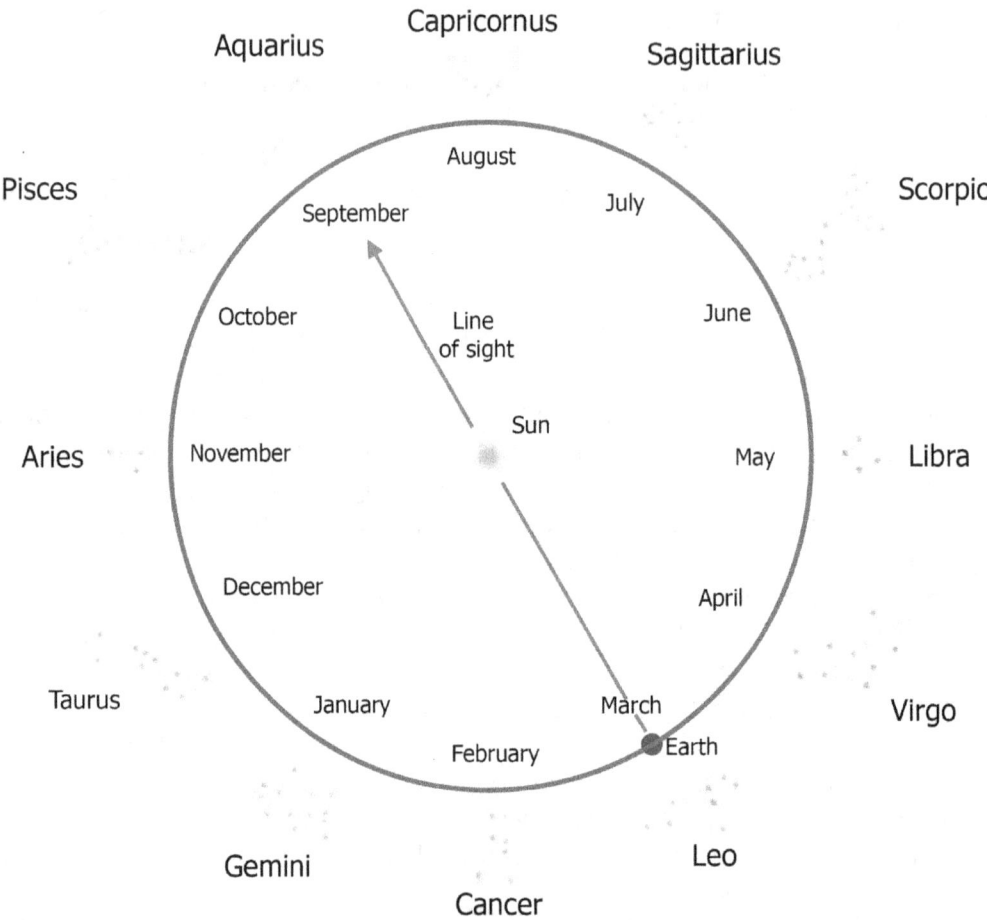

Position of the Sun and visibility of the zodiac constellations. Because of the Earth's orbital movements, the visible constellations change during the course of the year. The Ophiucus constellation, which is crossed between the end of November and the first week in December, between Scorpio and Sagittarius, isn't shown. The arrow indicates the direction in which the Sun will be projected in the chosen month. For example, the Sun will be projected toward Aquarius Constellation in March, so the constellations on the opposite side, like Leo, Virgo and Cancer will be more visible.

Finding the Planets

The planets are our neighbors, bodies much smaller than the stars, which circle in an orderly fashion around the Sun. They're always gigantic objects to us, and yet, compared to everything that's in the Universe, they seem to be tiny dots. Their strength is that since they are very close, on the scale of the Cosmo, they can be easily detected by the naked eye and, as we'll see further ahead, beautiful when observed with the telescope.

Which planets can we see with the naked eye? And how bright are they?

Let's begin in order by distance from the Sun, so we can also review a bit of theoretical astronomy.

Mercury, the most internal, is among the most difficult to see with the naked eye because it always shows up prospectively close to the Sun. At best, we have an hour after the Sun sets or before it rises, with an already clear sky. In addition to good eyesight, we have to understand where it is. The best times are called maximum elongations, when the planet is able to move away up to about 20 degrees from the Sun for a few days. To try and see it, we have to understand when these times are and in which area they are. The eastern elongations can be seen in the evening around half an hour after the sunset, low on the horizon in the direction of the Sun. The western elongations tell us that Mercury will be visible in the morning, about half an hour before the Sun rises, always in its direction. It's not absolutely faint, since it actually does reach a negative magnitude, but it's always immersed in the twilight, and therefore it's not easy to sight. I was able to see it only after years of attempts and, I must admit, a little by chance.

If someone wants to try, here are the best times to identify it.

Maximum elongations for Mercury in the next few years.

Date	When	Elongation	Magnitude
15 December 2018	Morning	22.7°	-0.1
27 February 2019	Evening	18.1	-0.2
11 April 2019	Morning	27.7°	+0.6
23 June 2019	Evening	25.2°	+0.7
9 August 2019	Morning	19.0°	+0.3
20 October 2019	Evening	24.6°	+0.1
28 November 2019	Morning	20.1	-0.3
10 February 2020	Evening	18.2°	-0.3
24 March 2020	Morning	27.8°	+0.5
4 June 2020	Evening	25.8°	+0.7
22 July 2020	Morning	20.1°	+0.5

Date	When	Elongation	Magnitude
1 October 2020	Evening	25.8°	+0.3
10 November 2020	Morning	19.1°	-0.3
24 January 2021	Evening	18.6°	-0.3
6 March 2021	Morning	27.3°	+0.4
17 May 2021	Evening	22.0°	+0.6
4 July 2021	Morning	21.6°	+0.6
14 September 2021	Evening	26.8°	+0.4
25 October 2021	Morning	18.4°	-0.3
7 January 2022	Evening	19.2°	-0.3
16 February 2022	Morning	26.3°	+0.2
29 April 2022	Evening	20.6°	+0.5
16 June 2022	Morning	23.2°	+0.7
27 August 2022	Evening	27.3°	+0.5
8 October 2022	Morning	18.0°	-0.3
21 December 2022	Evening	20.1°	-0.3

Venus, like Mercury, is an internal planet but it's much closer to the Earth, so it moves up to 48° away from the Sun, enough to be seen high in the sky after the sunset or before dawn. It shines with a magnitude equal to -4, and can reach -5 in the periods of its maximum proximity. This means that Venus is the brightest object of a stellar nature in the entire sky. Therefore, when it is present it's the first "star" to "turn on" after the Sun has gone below the horizon. Expert observers can even see it during the day, because its light is so intense that it never disappears in the brightness of the blue sky.

Even in this case the best times are during the maximum elongations, and if you don't want to spend months looking for it in vain, like I did, it's best to know that it can be seen before dawn or after the Sun sets, always in the direction of the Sun. Contrary to Mercury which is rarely visible, Venus can be observed for more than three months before and more than three months after the date of maximum distance from the Sun.

Maximum elongations of Venus in the next few years

Date	When	Elongation	Magnitude
16 January 2019	Morning	47.0°	-4.1
24 March 2020	Evening	46.1°	-4.1
13 August 2020	Morning	45.8°	-4.1
29 October 2021	Evening	47.0°	-4.2
20 March 2022	Morning	46.6°	-4.2
4 June 2023	Evening	45.4°	-4.1
23 October 2023	Morning	46.4°	-4.1

Mercury (bottom) and Venus (above), quite visible after sunset or before dawn, low on the horizon. Fuji compact digital camera (February 4, 2007).

Mars, the red planet; when it's present in the sky it shines with an unmistakable red/orange light that's impossible not to notice. It's the first external planet and the last rocky one in order of distance from the sun and can therefore be traced anywhere in the region called ecliptic that overlaps the zodiac constellations (do we remember that odd name?). If it's not too close to the Sun, it's always visible to the naked eye, but it noticeably changes brightness with the passing of time, going from a magnitude similar to that of the North Star all the way past Sirius. This depends on its distance from the Earth at the moment we observe it, which becomes minimum only every 26 months.

If we're patient enough and we like to draw, we can try to follow Mars' movements when it's near the Earth for a month or so, taking note of its position in regard to the stars every 2-3 days. We'll notice that it moves like all of the planets, but does so in a very strange manner: at times it seems to go ahead of the stars and other times in behind, drawing a curious 'S' in the sky. This strange movement is due to the fact that the Earth is also moving, with a revolution period that is around half of Mars'. When we see it in the sky we're actually observing the sum of the two movements which, at times, combine in a bizarre manner.

Oppositions of Mars in the next few years

Date	Constellation	Magnitude	Angular diameter
13 October 2020	Pisces	-2.6	22.4"*
8 December 2022	Taurus	-1.8	17.0"
16 January 2025	Gemini	-1.4	14.5"
19 February 2027	Leo	-1.4	14.5"

Mars (bright dot on the left) dominated the very dark sky of Atacama desert, Chile, during the 2018 great opposition.

Jupiter is the first of the gaseous giants, an enormous sphere, more than 11 times larger than the Earth, composed almost exclusively of gas. In average, it's 310,685,596.119 miles away but is visible for 10 months of the year and its brightness doesn't change much, remaining at a magnitude of around -2.5. White, it is always brighter than Sirius and is very easy to recognize in the sky: you just have to look for it.

---

* This strange symbol means arc seconds. I'll explain that further in the book, but we can try to discover the meaning of that before I'll unveil the mystery. Tip: it's an angular measurement.

Jupiter is about to set from the ESO VLT observatory located in Cerro Paranal, Atacama desert, Chile. Notice how bright is the center of the Milky Way in this 30 seconds single image taken on July 2018. At bottom we can see the Moon setting, generating colors similar to the sunset.

Instead, **Saturn** is the most difficult to track down because it shines with a magnitude of around 0, very similar to that of the star Vega. Its color is slightly yellowish and it moves very slowly in the sky, covering the entire path of the ecliptic in 30 years. However, even the most careful observer can miss the secret that it guards: a beautiful system of rings, which will, however, be an easy prey for our telescope.

Saturn (the brightest dot in the image) transiting in front of the center of the Milky Way. Image taken just after the sunset from the Atacama desert, Chile. Can you identify the upside-down constellations? Tip: rotate the book 90° counter-clockwise. October 2018.

Uranus is another gaseous giant, very far from the Earth, about 1864113577 miles away and is very difficult to track down because it shines at a 5.7 magnitude, close to the limit of the eye's possibilities. Its very slow movement in the sky (it takes 84 years to complete its circle around the ecliptic), as well as the difficulty in recognizing it in the middle of hundreds of stars, rendered it invisible to all the ancient populations, despite their having a lot of time to observe a pitch black sky.

The planets that are visible to the naked eye end here. Neptune and the ex-planet Pluto, now classified as a dwarf planet, are completely invisible. They're something we can see with a telescope, but it's best to not expect too much.

Searching for the planets

Although bright and almost all of them are easy to find, the planets have one big defect: they move around the sky. So, no one can report the positions of the planets in a book and when I was younger this was a problem, because internet didn't exist and cell phones, big and bulky, only did one thing, forgotten at this point: phone calls.

So, how can we find them and be successful in our third (or fourth) outing among the stars? How can we understand which ones can be observed?

We need updated maps, which we can create and print with those software, which I have mentioned more than once, that simulate the sky. At this point there are many resources, not like when I was younger and we had to judge by looks or buy updated monthly astronomy magazines which were nowhere to be found. Internet is full of sites that tell us where the planets are and which can be observed, and even our smartphones have apps available that can show us by pointing them out in the sky just like we did when looking for the constellations.

If, instead we like to do everything with our own efforts and maybe even challenge a friend to see who can find them first, here are a few suggestions that will make us more self-sufficient and less dependent on all this technology:

- The planets are always and only found in the zodiac constellations. We'll never find a planet in Ursa Major or in Orion. This is why we spent an entire night trying to identify them. If we know them well enough, we'll know right away how to identify a strange spot that shouldn't be there. It will most assuredly be a planet;
- The light from the planets doesn't flicker. In fact, we'll most likely have noticed how the stars, above all the brightest ones and the lowest ones on the horizon, seem to pulsate and even change color very quickly. No danger; they're not UFOs, but the effect of the atmosphere that disturbs their light. Instead, the planets don't undergo this effect and their light always appears perfectly still;
- When they are near the Earth, Jupiter and Mars shine brighter than Sirius. Therefore, if we see a bright, unmoving spot, this is assuredly a planet;
- Venus, other than being the brightest of all, is always found near the position of the Sun, as is Mercury. Therefore, it's difficult to make a mistake.

Before long, with a little experience, we'll not only be able to recognize all of the bright planets, but we'll be able to astound everyone by finding them in the sky in only a few seconds, at any time of year. And maybe your friends and acquaintances will also ask you a question that I am always asked: "How can you recognize the planets so quickly?" And I, a little shyly, answer, "How do you recognize a friend in the middle of a crowd of people? Simple, because you know them so well!" So, let's bring lots of good will, as well as the information we have found on internet regarding at least the planets that are visible during the evening. I can also give you some general advice. If we don't see planets in the early evening, it could be worth it to do something a little crazy that will become almost normal with time: get up a little before dawn and see if the morning sky has some surprise for us. It has always worked for me.

The rare surprises, the eclipses

With the end of our evening among the planets, and with the hope that we were able to identify a few, let's take a well-deserved rest before taking an important leap: begin using an astronomical instrument to help us learn about the sky with the naked eye, which should continue for at least another three nights without touching or desiring a telescope (difficult, right?). Meanwhile, let's distract ourselves from the struggles with practical astronomy by talking about phenomena that, unfortunately, we can't always observe in the sky, but which are at the height of spectacular when we see them with our own eyes.

We've already talked about the zodiacal light, but now it's time for the real highlight.

Eclipses are natural shows that don't need a dark sky, but are very rare, especially the loveliest of all: **the total solar eclipses.**

These take place when the Moon, in its voyage around the Earth, is found perfectly aligned with the Sun, to the point that it covers it perfectly. The total phase of a solar eclipse lasts, at the most, seven minutes; just a few, but among the most intense of our entire life. When the lunar disc covers the Sun perfectly, everything around us turns dark. The sky is dark near the Sun and bright on the horizon and the stars and planets become visible. But the most beautiful display is given to us by our star itself, because along with the blinding disc just covered the tenuous atmosphere, called the crown, becomes visible, extending for a few degrees and with an indescribable form, similar to that of the petals of a stupendous flower.

Unfortunately, total solar eclipses are very rare and even rarer to observe from our own home, because they take place over a strip of land several thousand miles long, but on average only about 125 miles wide. The last total solar eclipse visible in Italy (only by the central regions) took place in 1961 and the next will be in 2081!

So then, we have to accept moving, perhaps planning the trip far ahead and joining a big group of people lucky enough to have the time and means to go hunting down all the solar eclipses spread around the planet. It could seem like sheer madness, and I thought the same thing, too, before I saw one. Since then – it was November 14, 2012 – I've wondered why I waited so long to watch Nature's most impressive display, something that each of us, at least once in our life, should absolutely see.

Sometimes – even if we are in the right place at the right time – we might not see a total eclipse but instead one that is called a total-annular eclipse. During its path around the Earth, the Moon doesn't maintain a constant distance, but approaches and moves away for more than 25,000 miles during the arc of a rotation. If its alignment with the Earth and the Sun takes place near the maximum distance (called apogee) then the apparent diameter will be too small to entirely cover the solar disc. The Moon will enter the Sun and, in the final phase, it will seem as though there is a perfect ring of fire. It will seem beautiful and truthfully it is, but the effects of a total eclipse, such as the dark and the solar crown, won't be seen, necessarily rendering the annular eclipse a little less interesting than the total eclipse.

Solar Eclipses in upcoming years

Date	Type	Time of Maximum (UT)	Duration of Totality	Where it can be best seen
2 July 2019	Total	19:24	04m 33s	Argentina and Chile (total); South America, Central America and Polynesia (partial)
26 December 2019	Annular	05:18	03m40s	Asia, Australia
21 June 2020	Annular	06:41	00m38s	Africa, Asia, South-eastern Europe
14 December 2020	Total	16:14	02m10s	Argentina, Chile, Kiribati (total); Central and South America, Southwestern Africa (partial)
10 June 2021	Annular	10:43	05m51s	Northern Canada, eastern Russia, Greenland (annular), United States (partial)
4 December 2021	Total	07:34	01m54s	Antarctica (total), south America, South Africa (partial)
30 April 2022	Partial	20:42		Southeast Pacific, South America
25 October 2022	Partial	11:01		Europe, Northeast Africa
20 April 2023	Hybrid	04:18	01m16s	Western Australia, Indonesia. Papua New Guinea (total), China, Japan, India (partial)
14 October 2023	Annular	18:00	05m17s	United States, Central America (annular), South America (partial)
8 April 2024	Total	18:18	04m28s	Mexico, United States (total), Canada, central and south America (partial)

The impressive solar corona appears during a total solar eclipse. This photo is an attempt to represent – as much as possible – what the naked eye sees in those few minutes in which everything goes dark. Wyoming, August 21, 2017.

Lunar eclipses are more frequent and less spectacular but, fortunately, we can admire them without having to change continents. They take place when the Earth is between the Sun and the Moon, which is then darkened by the shadow produced by our planet.

They take place only when the Moon is full. In these circumstances, our satellite is completely eaten by the Earth's shadow in just a few dozen minutes. When the eclipse is total, the Moon shines with a weak, reddish light.

The famous red Moon is due to the fact that that the earth has a thick atmosphere that deviates the solar rays like a glass prism, until it illuminates the Moon despite it being in the middle of the Earth's shadow.

Lunar Eclipses in upcoming years

Date	Type of Eclipse	Maximum (UT)	Where can be seen
21 January 2019	Total	05:12	America, Europe, Africa
16 July 2019	Partial	21:31	Asia, Europe, Africa
26 May 2021	Total	11:19	East Asia, Oceania, West North America
19 November 2021	Partial	09:03	America
16 May 2022	Total	04:11	Europe, America
8 November 2022	Total	10:59	East Asia, Oceania, West North America

During the total phase of a lunar eclipse, our satellite turns a suggestive red color.

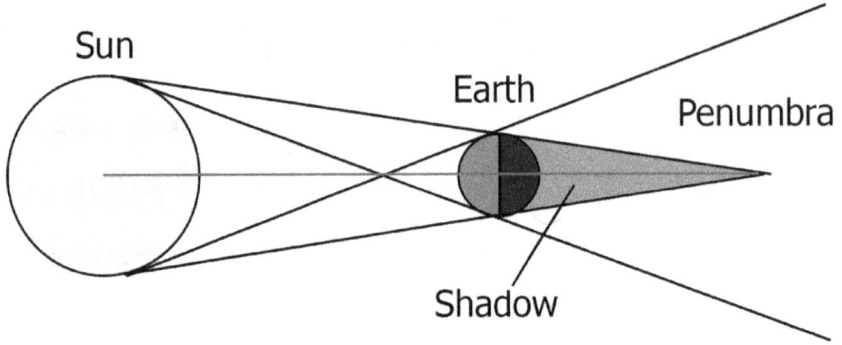

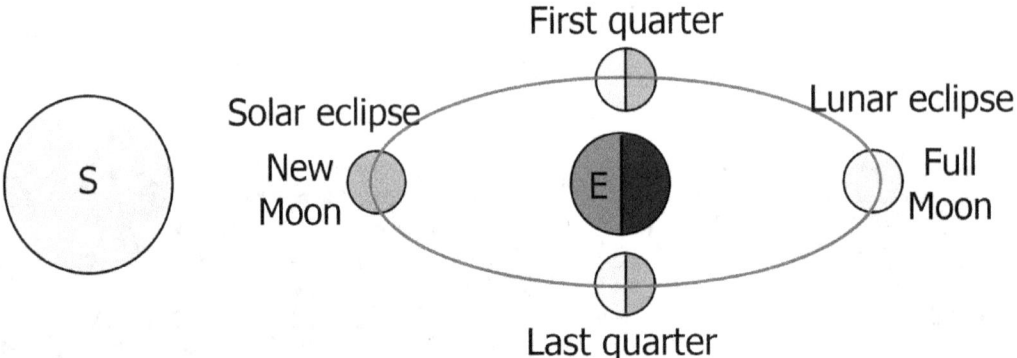

Lunar and solar eclipses are very rare and spectacular perfect cosmic alignments.

A help in the field: amateur astronomers and star parties

Many sky enthusiasts, especially in the new Moon periods, often get together and observe in localities that are particularly appropriate for astronomic observations. These group observations are called *star parties* and represent the best way to begin to know the sky in the company of those who have been watching for several years.

Help in the field from people who are prepared to guide observations by using their instruments represents the best way to learn the basics in amateur astronomy, to learn about telescopes, to understand what accessories we need, and to just exchange experiences and knowledge.

Don't be afraid to contact an astronomy club close to you or to reach enthusiasts during their outings.

Star parties are (almost) always free and full of interesting people, as well instruments that many (myself included) will probably never be able to have because of economic reasons or lack of space or time. The initiatives usually take place over the weekends, with a logical preference for the mild seasons (spring and late summer).

At night, the allure of a dark sky and dozens of enthusiasts ready to share emotions and experiences; during the day, conferences and activities (for example, observing the Sun) will present you with a truly unforgettable *full-immersion.*

The large *Dobson* telescopes, which open up to more than 20 inches, are always present in these manifestations and are so very rightly taken by storm by the amateur astronomers. I assure you that observing from these enormous instruments and finally seeing celestial objects similar to the photographs that we have seen on internet and in books and magazines, is worth standing in line for a little while.

Other than the big events, the tight network of amateur astronomers concentrated in your territory organizes events practically every weekend without clouds, defying sweltering heat, mosquitos, snow and ice. Some volunteers load their equipment in cars and cover hundreds of miles searching for a dark sky. Others venture to high mountains to enjoy the most stable and transparent skies possible.

If there is an astronomy club near you, participate in it and you'll learn in very little time, and certainly more directly, what you are reading in this volume: direct contact with the sky and the best way to know it.

Other than this, you'll understand that amateur astronomers are really very particular beings and, all things considered, fun and friendly!

The instruments are set up; everything is ready for the "star party" that will begin when the Sun sets. Observing in the company of other astrophiles is a great (and free!) way to share a common passion and to learn to know the sky, even if we don't have a telescope.

The sky through a binoculars

I did my first astronomic observation – even before I looked with my unaided eye – with those Russian binoculars that my father gifted me for my tenth birthday. I unwittingly aimed it at the Moon, and by now we all know what happened next.

I know that the desire to buy a telescope or to use one you already have is irresistible and that we have already taken great steps forward; therefore, I would never dream of telling you to buy binoculars at any price. However, if they are available to you (ask your parents or various relatives: at least someone will have a pair), using them represents a great training ground for gaining confidence with the enlarged sky and to have views that no telescope can ever give us because of a much too narrow view.

How are binoculars made and how do they work?

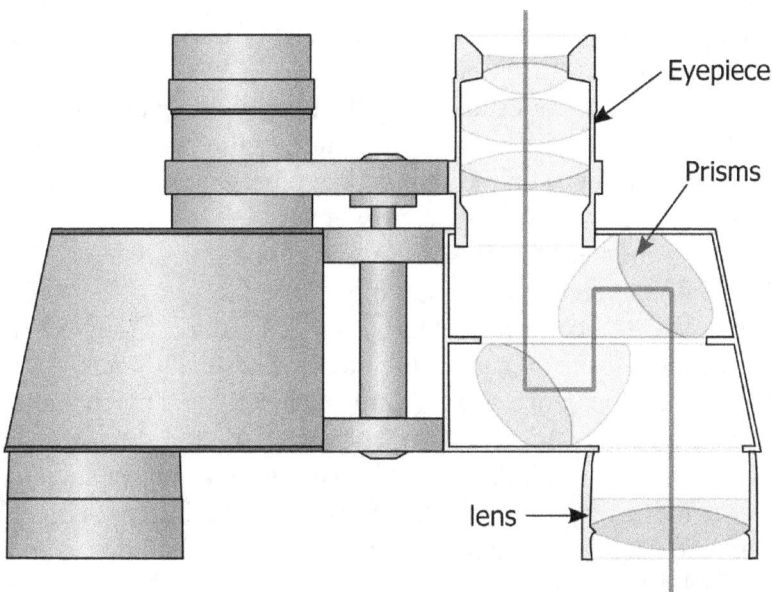

Scheme of a classical binoculars, formed by two objectives which channel the light toward prisms which have the task of straightening the image and bringing the optical axis together.

Binoculars are an optical instrument dedicated to terrestrial observations but can also be used profitably for observing the sky.

Binoculars are constituted by two objectives: the two larger openings, through which the light enters. Light is refracted and reflected by prisms that have the task of straightening the image that is naturally overturned by every optical system, until it merges into two small openings called eyepieces, or oculars, where we put our eyes.

All binoculars are characterized by two numbers that indicate the power: the diameter of the objectives and the magnification; they are usually imprinted on the body of the instrument. Binoculars on which the inscription 10X50 is engraved identify an instrument that magnifies 10 times and each objective has a diameter of 50mm (2 inches). Contrary to telescopes, the magnification can almost never be varied, and so it becomes, in this situation, an important evaluation factor. The optimal binoculars for beginning to observe the sky are the 7X50, 10X50 and 12X50. These instruments are both light and powerful at the same time, with a low enough magnification to be able to support them with your hand without the aid of an awkward tripod.

There are various types of binoculars on the market, from toy-like (like a 5X20) to large, cumbersome instruments, with typical telescopic diameters (20X80, 20X100). While the former won't provide any satisfaction in observing the sky, the latter are challenging instruments that require a heavy tripod for adequate support and are therefore unadvisable for amateurs.

The price for good binoculars is generally less than that for refracting telescopes (telescopes with lenses) of an equal diameter, since the optics manufacturing may not be perfect, but sufficient to support the enlargement for which it was planned. Despite this, it's best to be wary of seemingly advantageous offers: acceptable quality binoculars are rarely found for less than $50.00.

There are diverse methods for establishing the quality of a pair of binoculars, but the most important is, without a doubt, the collimation of the two objectives. Simply said, the two objectives must focus on the same zone; otherwise, the images will appear as overlapping to you, as though you were cross-eyed.

Many amateur astronomers underestimate the importance of the collimation of the objectives, but it's fundamental for observing comfortably.

In fact, a slight off-balance is corrected by our brain during the observation, which will, however, become tiring and annoying after just a few minutes.

Observing through binoculars that are in order optically and mechanically is relaxing and restful even after many minutes. If it's not, then try and see if the two objectives focus on exactly the same field. Place the binoculars on a support and look alternately with the left eye and the right eye: if you notice that the internal objects move, then your binoculars are slightly off-balance and this, in all probability, is the cause of the effort that you feel during your observations.

If the off-balance is serious, it will be impossible to have successful observations. If you are trying out the instrument, I advise you not to buy it. If, on the other hand, the binoculars are already yours, resist the temptation to throw it away, because in most cases the instrument can be refocused by working on the small screws on the sides intended for slightly moving the prisms. If you don't know how to do it, get advice from an expert and you'll see that the binoculars will return to a new and unexpected life.

The first attempt with your binoculars

Once you have binoculars available to you, I'd say we're at a good point. All of the efforts made in learning how to find our way about in the sky are now bearing fruit, because now we know what to do. Find a dark place, without the Moon, wait 15 minutes for your eyes to adapt to the dark and look through the instrument.

Right away, you'll notice that the difference between the naked eye and the instrument is enormous. Everything will move; you won't be able to find anything, not the even the smallest constellation will come into your field of vision. It almost seems like we can see better with the naked eye, but let's wait before we talk. Now that we know the sky we can head toward the most interesting zones, like the summer or winter Milky Way. If we're in a middle season, we can try to move around in Taurus or Cassiopeia if it's autumn, or in the weak Cancer constellation if it's spring.

After practicing a little (a few minutes are enough, but we need a lot of patience) we'll begin to see things that we couldn't see with the naked eye. Thousands of stars and shiny gobs will meet as they scud around the Milky Way, while chests of stars called open clusters will light up if we watch them in spring or fall in the zones that I suggested.

This is only the beginning; the spectacular impact is what we need to find the desire and the strength to continue, even more excited and determined than before. There's time to observe knowledgeably and understand what we are observing. Now, let's enjoy a healthy and pure cosmic display.

Preparing the first "official" evening with the binoculars

And here we are, a bit stunned after our first experience with binoculars; we're certainly excited, maybe even cold. Yes, just like what has often happened to me, in the most exciting moments, when I want to do something very much, I forget about everything else, especially heavy clothing. No problem: the important thing is that we had fun.

Now that we're back home, it's my job to take you by the hand and suggest what to do with the binoculars.

First of all, let's not forget the advice we followed for the naked eye observations: compass, red light, maps, clothes, maybe even a little company. But in this case, there's even more, because now, with a very narrow field of view, we need to be a little more careful, or else we'll get lost (sometimes it's okay but, maybe every once in a while, it's nice to know where we're going!).

The actual binocular observation must be prepared far more scrupulously than that with the naked eye also because, as we should have learned from our unbridled joyride, the objects to be observed change. If we can observe the constellations with the naked eye, we go beyond that with binoculars: star clusters, nebulae, galaxies... Knowing how to identify the figures in the sky will serve as a base for finding new objects.

So then, maps of the sky are essential for observing with binoculars.

The following figure represents an example prepared by the free program *SkyChart*, which depicts the *Lyra* constellation. It's easy to notice how the magnification is greater and how many more stars the program lets us see, many of which aren't visible to the naked eye. We're slowly going deeper into space, where very few have gone, and it's a beautiful sensation.

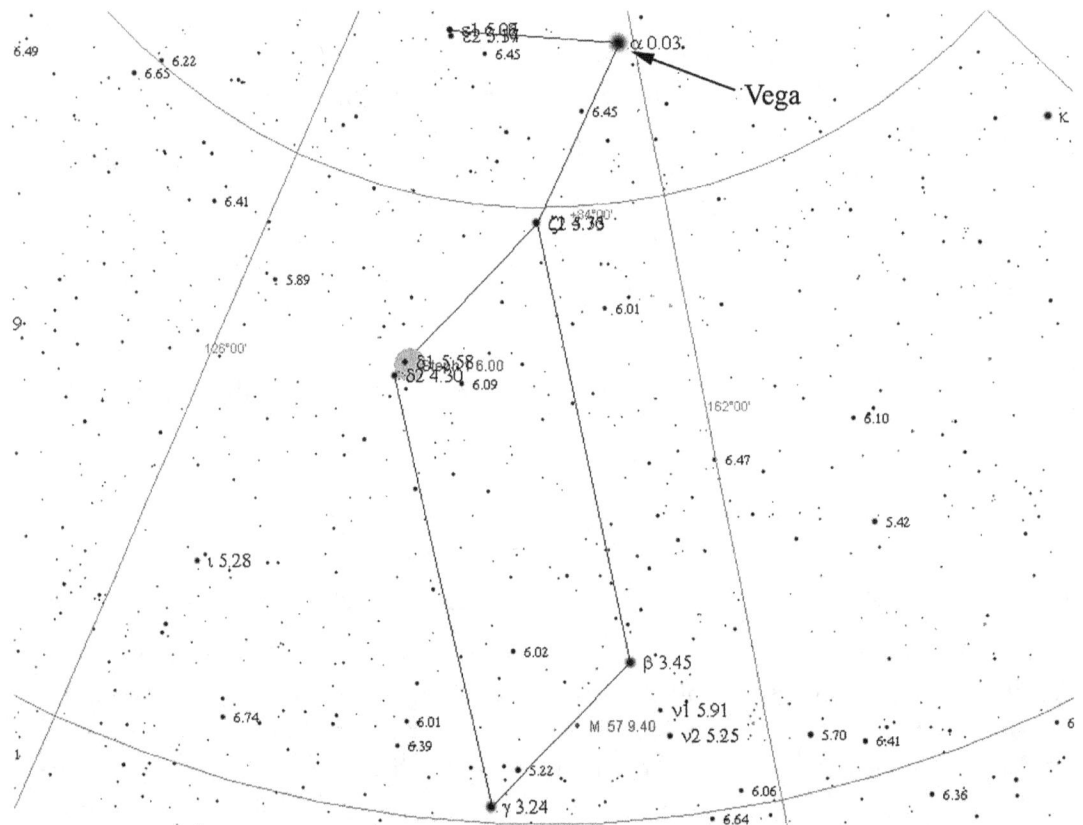

Example of a map for binocular and telescopic observations, prepared with the *SkyChart* software.

For our binoculars evening let's choose 5-6 constellations, possibly in the Milky Way disc (summer or winter), and let's get the software to show us the main objects and the magnitude of some weak stars between 5 and 7.

A few other non-astronomic features will help us appreciate the moment even more.

A lounge-chair that allows us to stay with our head skyward without the binoculars weighing us down will help us feel more comfortable during the observation.

With the naked eye, we learned that astronomy requires time and patience and that contemplation of the sky doesn't go well with haste. The latter is a hard and fast rule with binoculars and even more so with telescopes. The observation of the sky isn't appropriate for those who are in a hurry to see everything and yesterday. Rather, it's something that puts us in contact with the Universe, with our origins, with the important questions that everyone should ask at some time in their life. It has the taste of pure contemplation, of an intimate moment that we wish would never end, of tranquility and inner peace that only the sky has the potential of giving to us. Knowing how to capture all of this naturally depends on us. So let's get calm and relaxed and let the Universe guide all of our senses on an unparalleled voyage.

How to observe with binoculars

The maps are reference points for finding the objects in the celestial sphere to be observed. This time, no technological contraption can come to our rescue, and if up to this moment tablets and smartphones have been our only resource, we certainly won't be in good shape because we'll have to learn how to learn how to read and use very ancient material that is still useful: paper. A red flashlight and paper maps will avoid the problem of being blinded by LCD screens and will constitute the simplest way to go hunting for objects.

One thing that we'll feel the need for in the next outings – and certainly sooner or later when we observe with the telescope – is a notebook and a pencil to take down our impressions of all the fantastic things we'll be able to observe. But we'll have time to talk about this when we begin using a telescope.

To more easily track down objects, we need to understand how extensive the field that framed by the binoculars is. If we don't already know how, we can estimate by observing a portion of the sky where we know the angular dimensions: the usual stars in the Big Dipper are just what we need. The field of view usually oscillates from 5° to 7°.

How do we find objects on the map?

This is a very important question that I asked myself when I was trying to understand something in the sky, but instead of finding it written in a book I found it on my own…a few years later. How can we track down a star cluster or a nebula that we can't see with the naked sky? In these cases, it's not enough to know in which constellation the object is found. We could be perfect navigators but if we don't develop a new technique we have to patiently fathom the entire constellation and start observing a luminous ball for even more than half an hour. In fact, this is the method I used in the beginning, but 90% of my experiences have shown errors that shouldn't be made; even this case is no exception.

Star hopping is the technique that foresees aiming at the desired object by helping ourselves with close, bright stars, with small movements from star to star. It's no surprise then that this expression tells us to hop from star to star. Let's remember it, because it'll help us a lot even, and especially when we'll be observing with a telescope.

So, this is why we should know the bright constellations fairly well, knowing how to estimate angular distances and the magnitudes of the stars.

Truthfully, we're not doing anything new because it is the same technique that we have unwittingly applied in finding the constellations, starting from already known figures, like the Big Dipper. The difference now is in making smaller movements and having a good enough precision to track down the object with binoculars.

A classic example of binocular or telescopic star hopping regards, for example, tracking down the M13 globular cluster, located in the Hercules constellation (have we already found it? If not, we can start from the brighter Vega), along the imaginary line that joins the η and ξ stars at about 1/3 of the distance that separates ξ from η. Oh dear, it seems complicated. And what are those strange symbols we just saw? Let's stay calm; they are actually very simple things. Other than a few bright stars that also have their own names, the stars in a constellation are called by the symbols of the Greek alphabet. The two that we have just seen are the letters xi (ξ) and eta (η). The letters are assigned from the brightest star to the weakest. So, the most luminous star in each constellation is named alpha (α), the second will be beta (β) and so forth. Each paper and digital map will give this nomenclature, which serves to identify the stars in every constellation. We don't even have to learn the Greek alphabet because it is already written at the beginning of this book (and maybe we didn't even notice it).

Once the two stars have been found on the map we'll try to identify, with the naked eye, the one closest to our objective; then we'll visualize in our mind the path to follow and the presumed position in which we have to head, in this case M13, and then, for last we'll take the instrument to calmly follow the same path that we have just traced. It's not really easy, but there's no one coming to hurry us along. With patience and determination we can do it, especially if we have precise maps like the ones on the following pages, and each time we do it, it will be easier.

So, searching for objects by star hopping becomes part of the excitement of an observation, a real exploration of the sky that puts patience and ability to the test. And then, when we finally find, through our own efforts and not Google's, the little ball that we've searched for, for so long, we'll feel a satisfaction we've rarely felt in our life. Yes, we have begun to explore the cosmos with just our efforts. Just us and the stars; a human being and the entire infinite firmament over our head.

A few other useful techniques

When following a great passion and an incredible desire to explore the sky, it seems that we have to do everything right now, as though someone were chasing us.

Several years have gone by but I still remember very well the craving to find an object and the frustration that already began just a few seconds of fruitless searching. I know it isn't easy to hold ourselves in check, but we have to do it. We have learned so many notions in so little time and our brain needs a little while longer to learn how to take advantage of them in the best possible way.

Having said this, I have learned a few techniques through experience that could be useful in speeding up the learning process. Besides being calm and patient, the need to never become discouraged and to continue trying, there is a very powerful technique for improving our view of objects, called averted vision.

The center of our eye – the part that we always use to see things (and to read these words) – is about 10 times less sensitive than the peripheral portion. Perhaps we weren't aware of it, but the effect is very evident at night on faint and perhaps hazy objects, like the Andromeda galaxy. Under a dark, Moonless sky, we have seen that it is perceptible to the naked eye, but seems strange.

If we indeed try to observe it directly, with the center of our eye, it seems to almost disappear. Then, when we're lost and start over from the beginning, starting from the nearest stars, we see it right away from the corner of our eye, brighter and more obvious than before. Then our instinct tells us to look at it directly again and it seems to suddenly disappear again. We are experimenting with the power of averted vision: if we observe something with the corner of our eye, focusing the center of our gaze toward a nearby star, we'll be able to see it brighter, with greater details.

The averted vision technique requires some training to be profitably put into practice because, at first, our eye will try at any cost to look directly at the object again but, with time, we'll know how to control it successfully.

Another technique for appreciating at best a celestial object consists in dedicating at least several minutes to it. At the beginning everything, except maybe the Moon, will seem small, indistinct, dim and disappointing, but most of the time it's an illusion. Our eye and brain are used to admiring the very bright daylight scenes or beautiful photographs that are full of details and contrasts. Nocturnal observations are very different and are a taste of what we have just seen, both practically and theoretically. The details tend to come out only after a couple of minutes and that apparently anonymous object slowly fills in with faint structures and shades. It's useless to complain if it still doesn't seem spectacular to us: this is the true Universe, whether we like it or not and if we can't appreciate it, maybe it's time to think about changing hobbies or to do it differently. I know many enthusiasts who limit themselves to photographing it because they want to literally see contrasts and colors literally jump from the computer screen. But photography isn't the subject of this book, because that subject is fairly complicated and expensive.

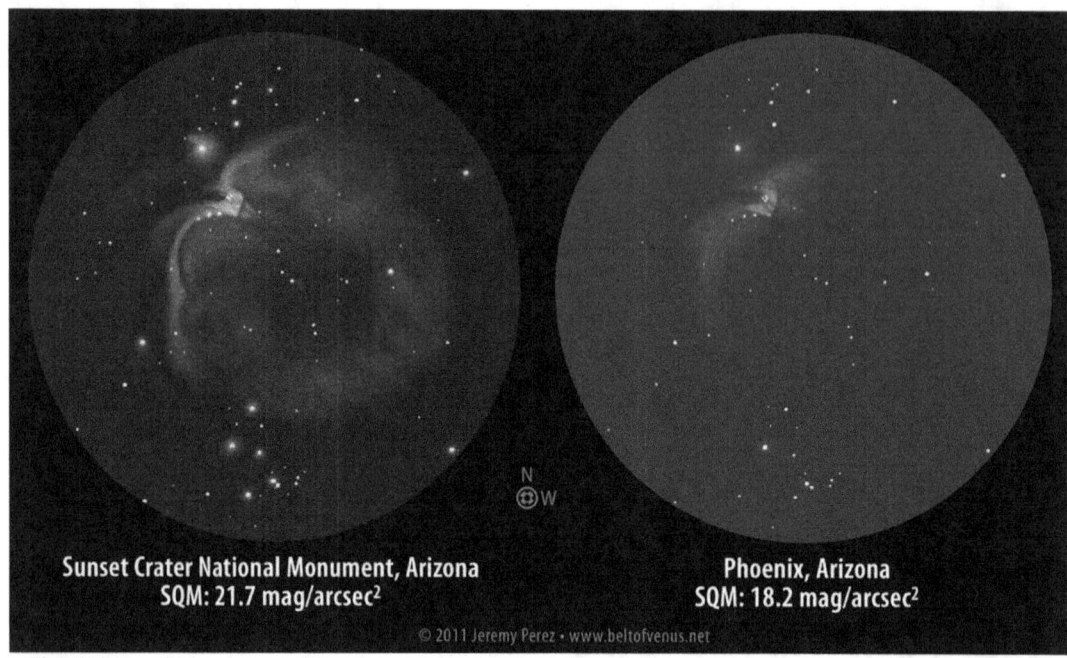

Sunset Crater National Monument, Arizona
SQM: 21.7 mag/arcsec^2

Phoenix, Arizona
SQM: 18.2 mag/arcsec^2

© 2011 Jeremy Perez • www.beltofvenus.net

The Orion nebula observed in a very dark sky (at left) and from a rather polluted sky (at right). The quality of the sky is fundamental for every deep-sky observation. The differences in these two sketches, however, could also represent an experienced observer (at left) versus a novice one (right), under the same sky, with the same telescope.

The Andromeda Galaxy observed using binoculars with at least 50mm (2 inches) lens is very beautiful and is accompanied by two small, dim balls: they are M32, of a star-like appearance and the closest to the nucleus, and M110, further away and widespread, two galaxies, satellites of this giant of the skies.

The summer Milky Way is great fun with binoculars. In this drawing, the Laguna Nebula, center, is visible even with the naked eye and, at above right, a weak corona characterizes the dim Trifid Nebula. No telescope will show them together as can be done by good binoculars.

Where is north and south?

Degrees and cardinal points are usually used to find celestial objects when using the star hopping technique.

The M92 globular cluster, for example, visible with small binoculars, is tracked down by moving 4° to the south west of the t Herculis star. It seems simple in theory, but there's a trap lurking that we have to keep well in mind, to avoid falling into it.

I no longer know how much time I lost when, as a young enthusiast with maps and astrolabe, I tried to point out objects with a telescope. I also had a compass, I knew how to move, but my manner was wrong. I thought that north was always up, south was always down, west was to the left and east was to the right, as though I was looking at a geographical map. But the celestial sphere isn't a planisphere.

We can't confuse the north with the direction up, the south with down and the east and the west with right and left. The directions in which we have to move are always referred to the cardinal points identified on the horizon, therefore set. If the constellation is found in the meridian and we're looking southward, the cardinal points of the horizon will coincide with the high-low and right-left direction. Instead, in all other cases, the cardinal points will be "distorted". So, going northward hardly ever means moving upward, just like going southward doesn't imply going downward, as we see on a geographical map.

If a constellation is found toward the eastern horizon because it has just risen, the north will be inclined and will be the line that joins the constellation with the point on the horizon where north is found (exactly below the North Star), south on the point of the opposite horizon, east, a little lower and west behind us. Maybe it's not easy to visualize, but with a little practice we'll avoid many stupid errors.

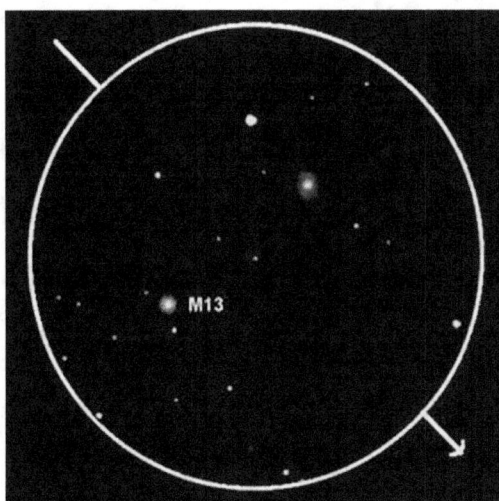
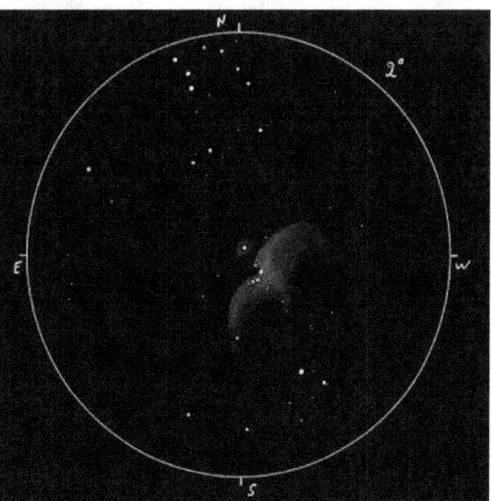

The M13 globular cluster in Hercules, as seen through 7X50 binoculars. The arrow indicates where the northern cardinal point is found.

The great Orion Nebula seen through a 10X50 binoculars. Notice the image's orientation compared to the cardinal points on our horizon.

Our first "official" observation

Good; everything is ready and I'll bet that we are psyched about our first "official" outing under the stars with binoculars. I am, while writing down these lines and, am reliving in my mind the unique sensations that one feels when getting ready to leave on a voyage through the Universe. We can't do it with spaceships like those in star Trek, but our instruments are just as powerful and let us see things that many humans have never seen during their entire life.

All things considered, we are privileged, people who have the chance to observe a much more ample and true reality than what human beings have created. The humans that arrived at the point of erasing the stars in the sky to be sure of not having to ask questions that they wouldn't have known how to answer. Because the unknown forces them to think and thinking often forces us to realize how stupid and superficial many of the actions we commonly do are. And this, ultimately, would force us to change, but change is frightening, so it's best to not be tempted to think and live in one's own fantasies. But not us; were not afraid of the unknown, we love the darkness because lets us to see the stars. We like to explore; we like living in the heart of the reality that is right above our heads. This society and this small world are too tight-fitting for us, because we have understood that our home is the Universe, a much different environment which has so much to teach us that will improve us day after day. We'll discover it very soon. For now, let's just enjoy our new adventure under the stars in a completely personal manner.

Binocular objects in autumn

Name	Constellation	Description
Double Cluster (NGC869-884)	Perseus (Per)	Pair of spectacular open clusters. They are found in the heart of the Milky Way between Cassiopeia and Perseus, even with the naked eye, as two small spots. Spectacular with binoculars, immersed in the Galaxy's rich star field.
M34	Perseus (Per)	Open cluster halfway between γ *Andromedae* and β *Persei (Algol)*. It appears partially nebulous with binoculars, with 5-6 easily seen stars, immersed in a round, extended spot.
M31 (Andromeda Galaxy)	Andromeda (And)	The well-known Andromeda galaxy; easily identified with the naked eye, beautiful with binoculars: more so than with a telescope. Located above β *Andromedae*, it appears oval and very bright. It is much more extensive when observed with attention and with adverted vision.

M15	Pegasus (Peg)	Globular cluster situated 4° south of ε, the southernmost star of the Pegasus Constellation.
M52	Cassiopeia (Cas)	Open cluster, easy to find, just outside the W of Cassiopeia.
M33	Triangle (Tri)	The well-known Triangle Galaxy is the next-to-the-last from the Earth and is barely visible to the naked eye in perfect skies (I saw it wonderfully in Australia!). Its fame isn't due to its beauty, but to the difficulty with which it is seen, even though it is, theoretically, bright. If the sky is very dark it's evident with binoculars as an indistinct cloud, not too far from M31.

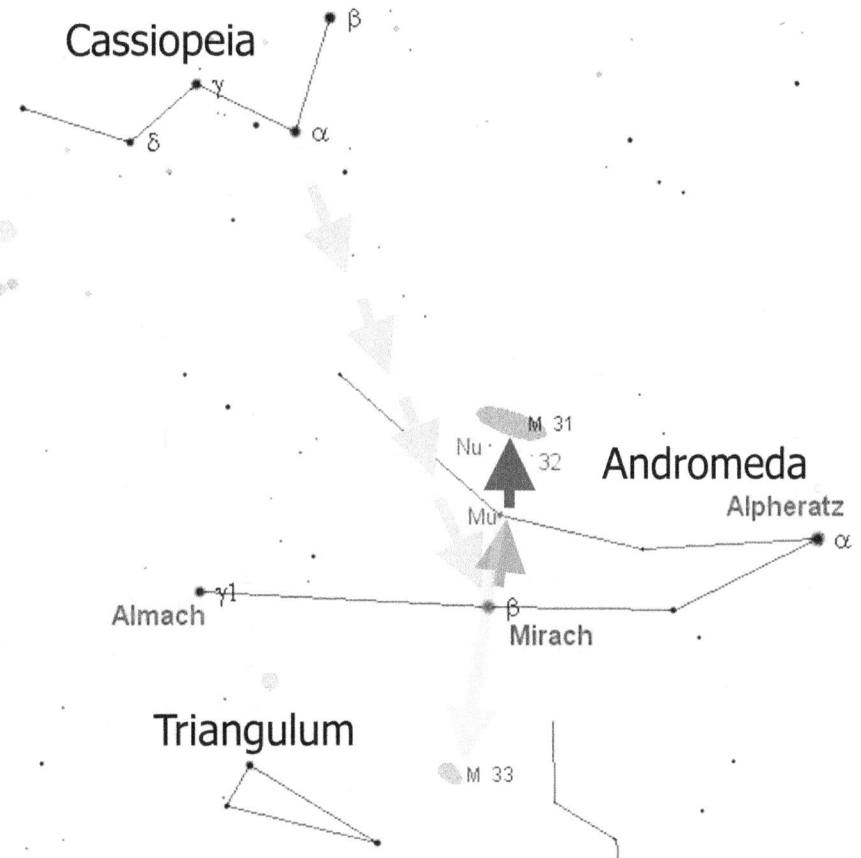

Star hopping to find the Andromeda Galaxy (M31) and another, slightly elusive galaxy: M33. Visible with binoculars, both appear as small indistinct clouds. However, be careful not to make a mistake: those soft, dim threads are actually hundreds of billions of stars and planets, perhaps some of which are inhabited, too close to each other and too far away from us to be seen. It seems impossible, but we are observing another island of stars, another real Universe, if we think that all the stars, nebulae and clusters visible in our sky belong to only one galaxy: the Milky Way.

Cassiopeia

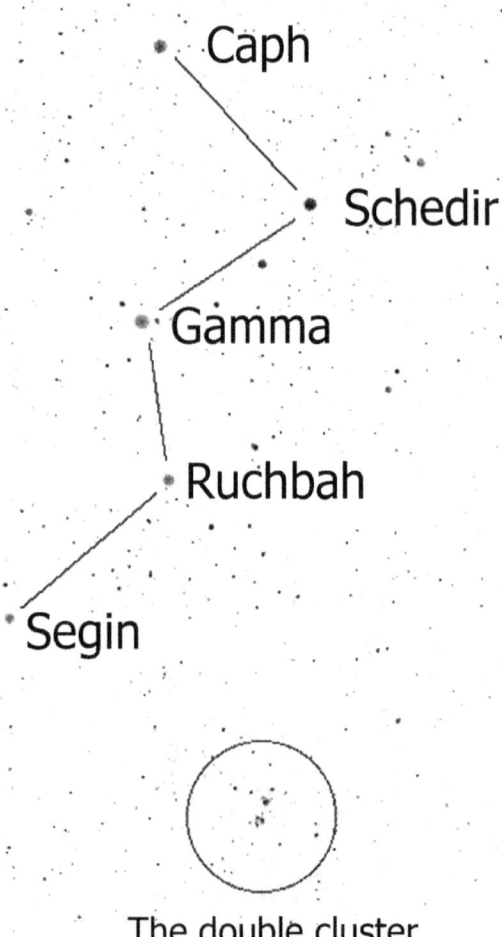

The double cluster of Perseus is visible even with the naked eye, not far from Cassiopeia, as a dim and indistinct cloud. With binoculars of at least 50mm (2 inches) it will be revealed as hundreds of tiny stars.

Messier 15 - M15

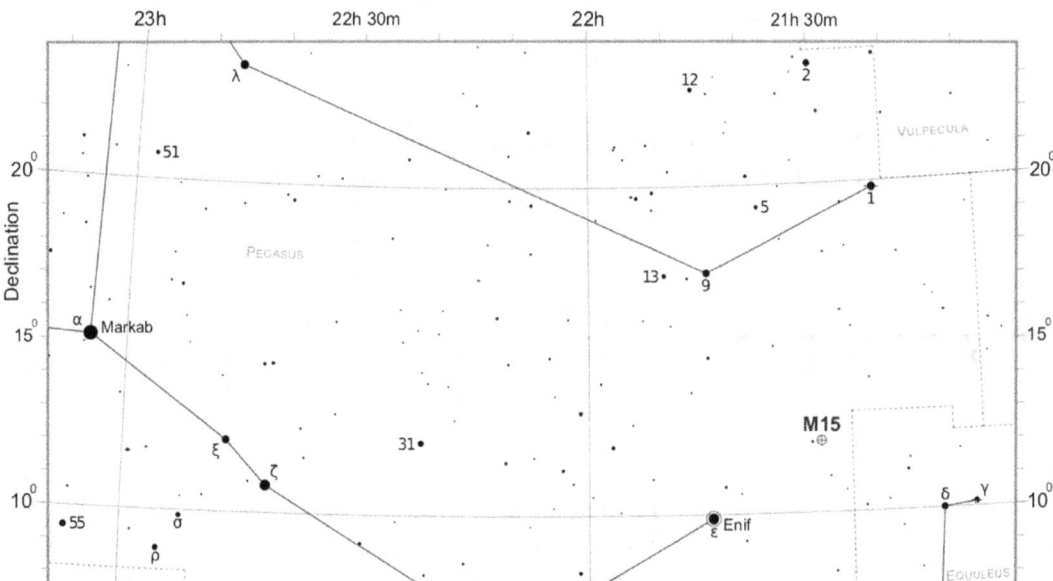

M15 is a globular cluster – not exactly easily found – located in the eastern portion of the Pegasus constellation. With binoculars, it appears as an unfocused star, but here, too, its appearance is deceptive: In fact, hundreds of thousands of stars are concentrated inside it. One wonders how they can all fit together in such a small space. Actually, it appears so tiny only to us, who are tens of thousands of light years away. Chart downloaded from https://freestarcharts.com

Binocular objects in winter

Name	Constellation	Description
Pleiades (M45)	Taurus (Tau)	Very famous open cluster, easy to observe with the naked eye even with polluted skies. Splendid when observed with any size binoculars.
Hyades	Taurus (Tau)	Very extensive open cluster, south of the Pleiades. It forms a sort of V with *Aldebaran* at the top, although the latter isn't a part of the cluster. It's one of the nearest open clusters to us, visible to the naked eye and very beautiful with binoculars.
M42	Orion (Ori)	The great Orion nebula doesn't need to be presented. It's very easy to find below Orion's Belt, half-way down the sword. To the naked eye, it appears as a blurry star; with binoculars, it is stupendous. In averted vision, soft shadings can be observed in the nebula's intricate weft, at the heart of which we can find the trapezium, a small open cluster.
M41	Canis Major (CMa)	Open cluster situated at 4° south of *Sirius*. As extended as the full Moon, it is easy to find and observe with binoculars, presenting a discreet concentration of bright stars. It's not visible to the naked eye.
M36 M37 M38	Auriga (Aur)	Three open clusters, all 3 clearly visible in the lower part of the *Auriga*. M36 is the smallest and brightest. M37 remains cloudy, while M38 resolves into single stars. This entire region of the sky is stupendous if observed with binoculars.
M35	Gemini (Gem)	Open cluster south of ε *Gemini*. Visible to the naked eye with an extremely dark sky; it's splendid with binoculars, which shows the individual stars.
M46 M47	Puppis (Pup)	Open clusters situated circa 8° north of *Sirius*, and so fairly easy to find. They are visible in the same field with binoculars. M47 is brighter and more resolved in stars; M46 is only partially resolvable with 50mm (2 inches) binoculars.

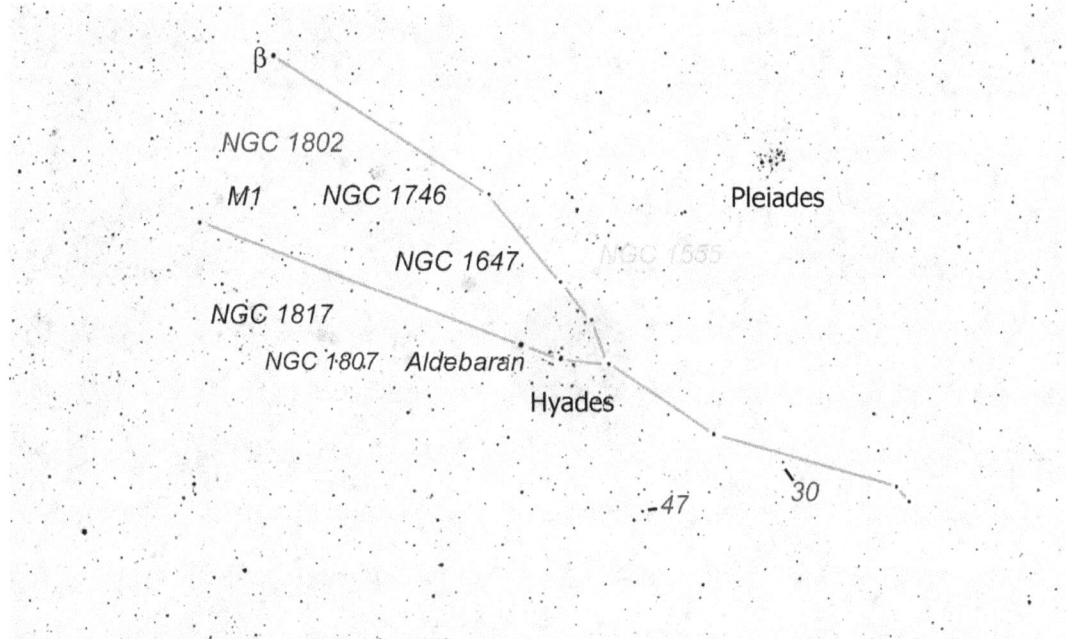

The Pleiades, also named M45, resemble a small dipper and are easily seen by the naked eye. Under a perfect sky – like that of the Atacama desert – I could count 12 stars without any instrument. From a normal dark sky, is it possible to see 7-8 components, but they become hundreds with a binoculars.

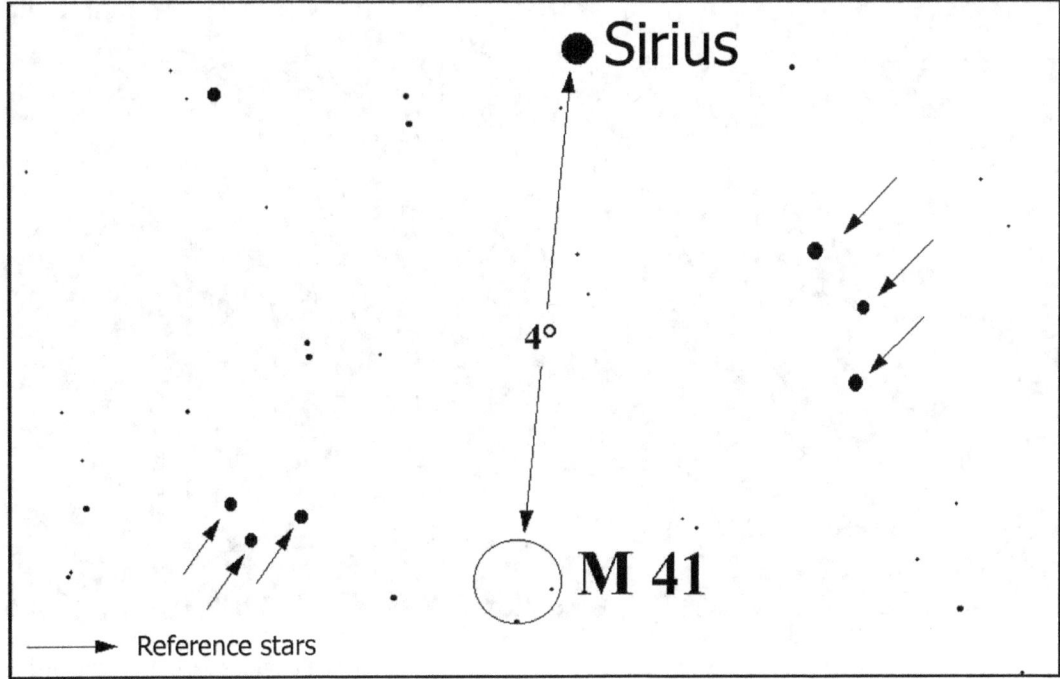

M41 is an open cluster situated 4° south of Sirius, the brightest star of the sky. It is seen very clearly with binoculars.

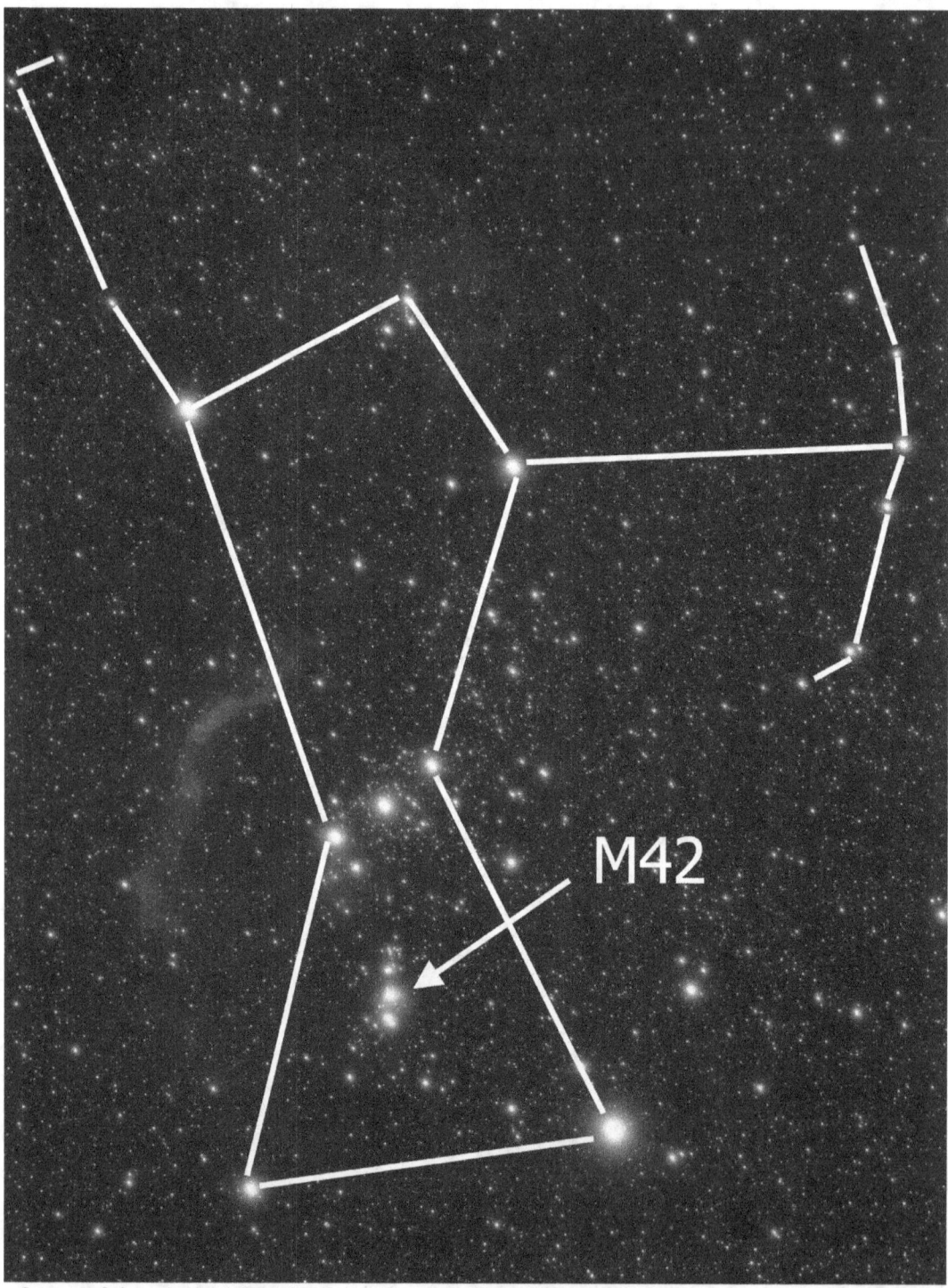

Queen of the winter nights, the Orion Constellation, with the Great Orion Nebula (M42), can't be missing from any astronomy buff's loot. A dark Moonless sky will show a tenuous gaseous cloud inside of which thousands of stars are being born.

Binocular objects in the spring

Name	Constellation	Description
M44	Cancer (Cnc)	The famous Praesepe open cluster (aka Beehive cluster) is easy to spot with the naked eye, if the sky is dark enough, located in the dim Cancer Constellation, of which it seems to be the most evident object, even more than the stars in the constellation. It was the first object observed through a telescope by Galileo in the seventeenth century, at the beginning of astronomic observations. It's spectacular with binoculars.
M67	Cancer (Cnc)	Another open cluster, very different regarding its concentration and age compared to M44. It's about 2° east of the α star. It's not easy to resolve with binoculars, except partially when using averted vision, .
M3	Canes Venatici (Cnv)	Globular cluster, easy to observe as a small, blurry white spot. It's not easy to identify because the zone has few stars; found at about 12° northeast of *Arcturus*.

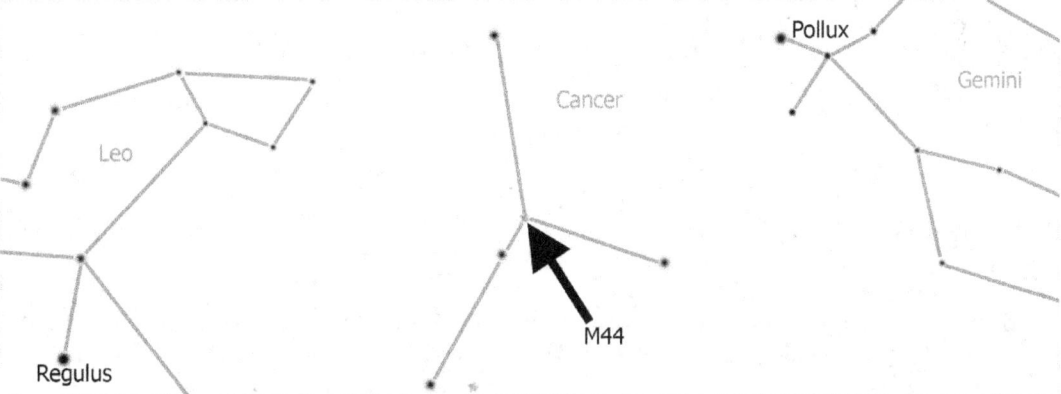

The most evident object in the faint Cancer constellation, perhaps more than the faint stars, M44 is visible to the naked eye as a milky blurry ball. It begins to show the stars with binoculars. In the same constellation, down lower, M67 can also be observed, but it's not visible to the naked eye. This is one of the oldest known open clusters.

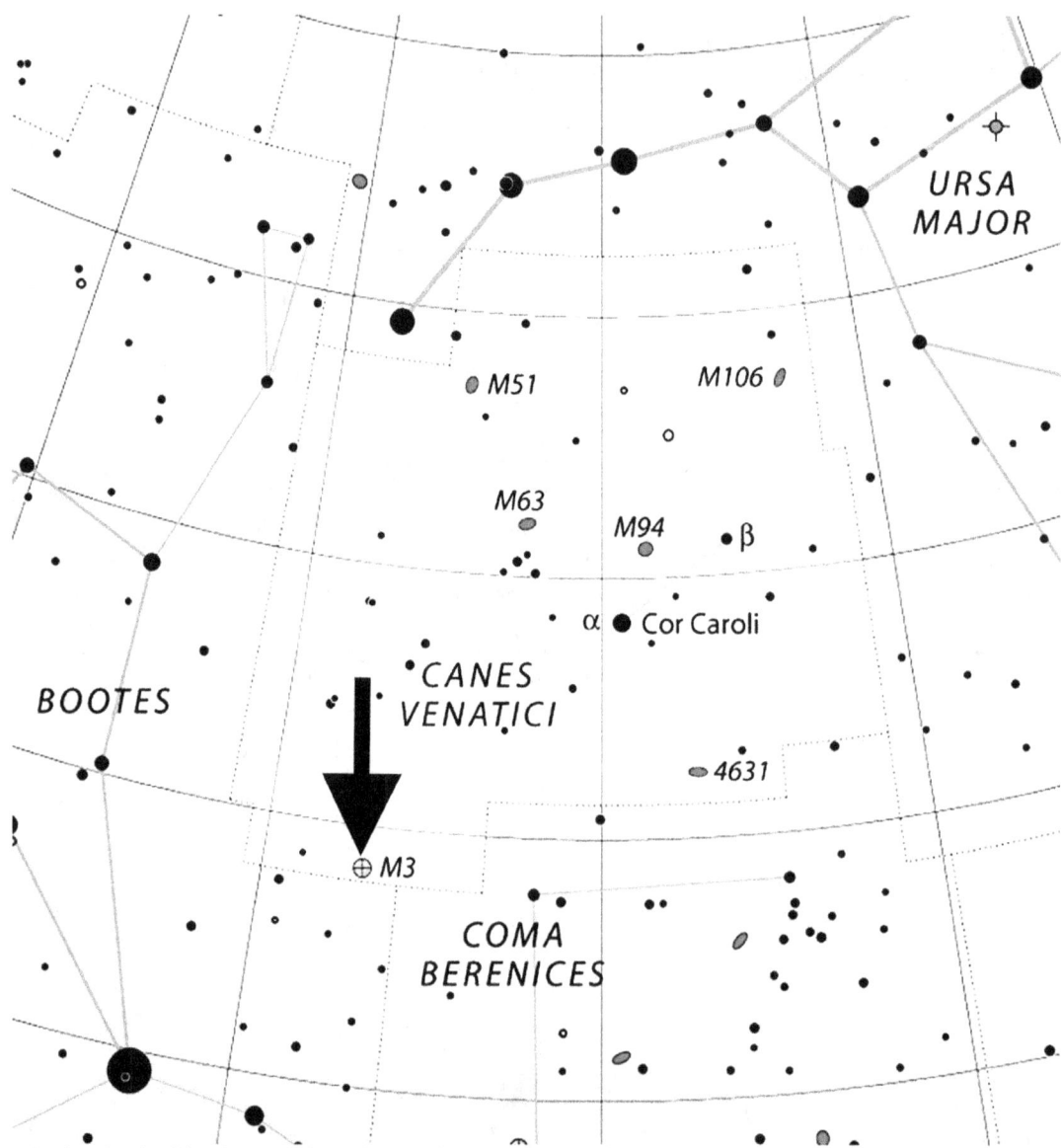

Spring is the kingdom of the galaxies, but a telescope is necessary for those. But, every once in a while, some brighter object appears, like the globular cluster M3.

Binocular objects in summer

Name	Constellation	Description
M13	Hercules (Her)	Famous globular cluster, visible even with the naked eye. It appears as a bright, round spot, traceable along the line that joins the η and ξ stars of the constellation, at about 1/3 the distance from ξ to η.
M92	Hercules (Her)	Globular cluster, slightly difficult to track down, because it is in an area with few stars. We can find it 4° south-west of τ *Herculis*. Quite beautiful with binoculars, with a cloudy aspect; a bit dim.
M4	Scorpio (Sco)	Another globular mass in the same field as *Antares*, the bright star of Scorpio. Weak, but rather extended.
M6 M7	Scorpio (Sco)	Open clusters in the Scorpion's tail, visible in the same field of view with any binoculars and perfectly resolved. A very suggestive view.
M22	Sagittarius (Sgr)	The brightest globular cluster visible in the borealis sky, it appears as a blurry star, with dimensions similar to those of the full Moon.
M23 M24	Sagittarius (Sgr)	M23 is a small open cluster, while M24 is a gigantic stellar full of colored stars, perfectly visible even with the naked eye. Absolutely wonderful with any binoculars.
M8 M17 M20	Sagittarius (Sgr)	Emission nebulae. M8 and M20 are very close to each other and visible in the same field. M8, called the Lagoon nebula, is perfectly visible even with the naked eye as a small cloud along the disc of the Milky Way. M20 is smaller and dim. M17 is found around a dozen degrees further north. It's called the Omega or Swan nebula, because of its characteristic shape and is perceptible only with binoculars.
M11	Scutum (Sct)	Among the most beautiful open clusters visible with binoculars, called the "Wild Duck". It's found at about 3° west of the λ star of the Eagle, located on the lower border of the constellation. It's extensive and bright; it begins to resolve with 80mm (3.15 inches) binoculars on very transparent nights.

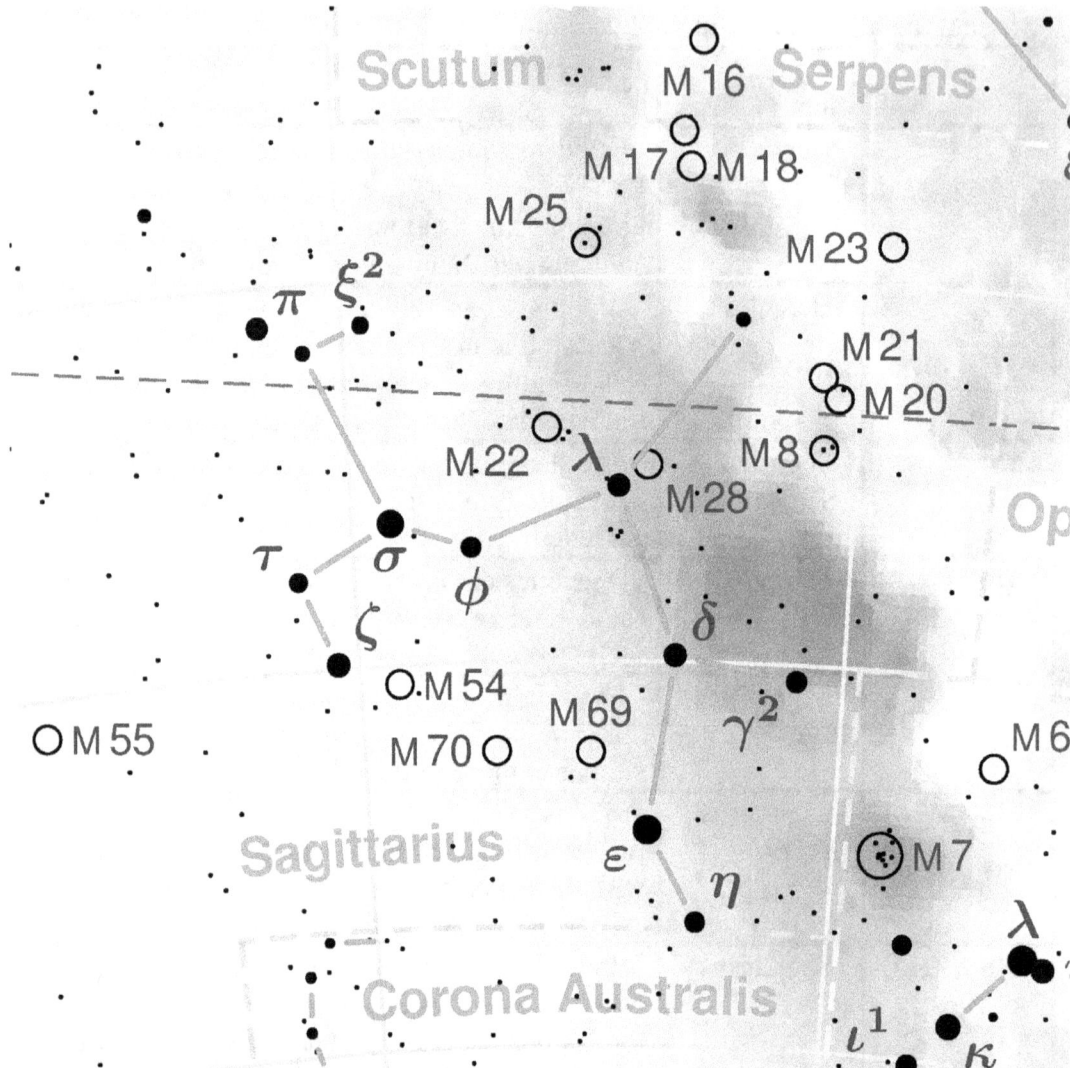

The summer Milky Way, as we have already been able to experience, is the best place to aim our binoculars in the discovery of thousands of stars, dozens of nebulae and open clusters. The Sagittarius constellation, overlapping the Galaxy's central zone, is particularly rich with stars: all we have to do is choose!

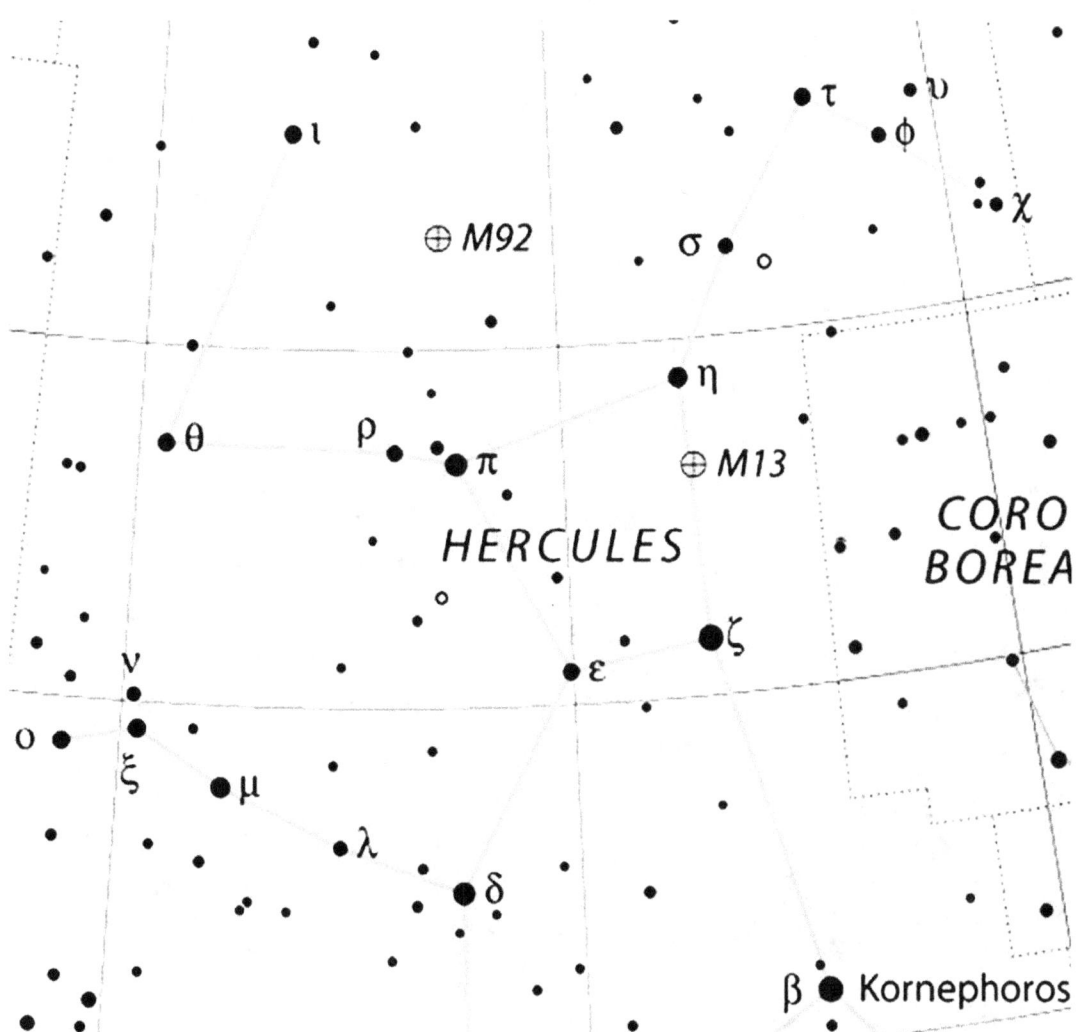

M13 and M92 are two very different bright globular clusters. M92 is small and concentrated, while M13 is much more extensive. Beautiful to observe with binoculars even though, like all globular clusters, they are best with a telescope.

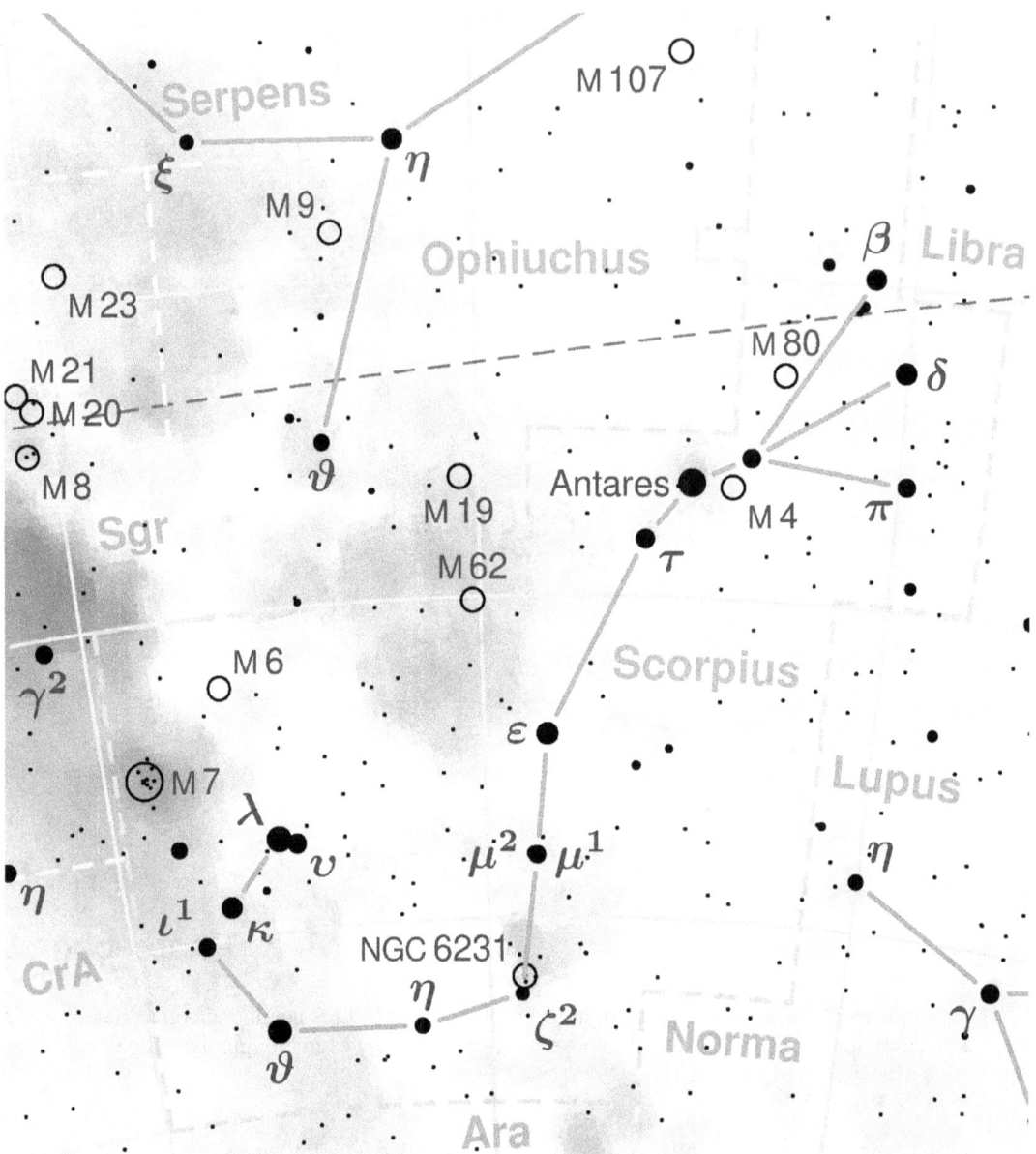

The Scorpius Constellation, located near the center of the Milky Way, is also full of binocular objects. If it's visible on the horizon, it certainly deserves a glance.

The Telescope

Universe, here we come, with our first telescope!

The long-awaited moment has finally come.

We've covered a long stretch of road since we first started. We couldn't even find the Northern Star; we didn't know what there was to observe in the sky, or even how to do it. Instead, in just a few days we have made great leaps. We've been good in following all advice, in not cutting corners, to be patient and determined, especially when everything seemed terribly difficult.

Now, we will be paid back for our efforts because it's time to talk about telescopes and understanding how to use them. The hard part is over, but never relax. Observing the sky should be done continuously because forgetting the constellations is much easier than you would think.

Let's never make the mistake of feeling that you've reached your goal and that all is well, because we never finish learning and maybe this is the best part of the game.

What is a telescope?

The telescope is the instrument dedicated to observing the sky par excellence. Created during Galileo Galilei's days (although he didn't invent it), it revolutionized astronomy thanks to the Pisano genius.

Now, more than 400 years later, this invention still continues to enjoy excellent health and doesn't seem to be headed toward its demise (luckily!).

Every telescope is formed by a group of lenses, mirrors or both which have the task of catching the light and concentrating it on a point called focus or focal point. The distance between the objective

A classical telescope, called refractor.

– the group that focuses the light – and the point where it is concentrated is called focal length. The functioning principle is the same as our eye, where the crystalline lens catches the light and directs it to the retina, the zone where the image that we can see is formed. In fact, all telescopes produce a visible image of what they have pointed onto the focal plane, just like a magnifying glass on a white sheet of paper (try it to believe it!).

Why does a telescope let us see better?

Each telescope, but also the binoculars we just saw, lets us see objects that the eye wouldn't be able to see. But, why? You could think it's all a question of magnification... Actually, this is what I thought when I was little. My first telescope didn't let me reach high magnifications, so I innocently tried to overlap every type of lens, trying to see better. In fact, after years of frustration, thinking that excessive magnification would let me see more, I decided to buy a second telescope that would allow me to do so. How did it go? I finally understood, the hard way, what a telescope's true power was and magnification's role.

The telescope's power, if it has been made well, is all in the diameter of its lens which is expressed in millimeters, just like it is with binoculars.

The diameter of the lenses, or mirrors, determines all of the telescope's characteristics: resolving power and the capacity for catching the light.

To say it more simply, a telescope of a certain diameter allows us to see details that are very small or very close to each other, and shows very faint objects.

Magnification has nothing to do with any of this, also because, theoretically, every telescope can reach the magnifications we desire.

A telescope's maximum performance, determined only by the diameter, is reached with a dark sky (the maximum capacity of catching the light) and with a maximum – a total maximum – magnification, is 2.5 times the diameter of the lens expressed in millimeters. So, an instrument with a 100 millimeter diameter allows us to enjoy magnifications that reach, in the best possible hypothesis, 250 times (usually written as 250X, read as 250 times). If we zoom in more than this, and no one will stop us, we'll begin to see hazy and increasingly dimmer images, losing the maximum power that we have with what is called maximum useful magnification. If we want to profitably use higher magnifications, we have to buy a telescope with a greater diameter, because otherwise we'll never see new celestial objects.

The capacity to see faint objects doesn't depend on magnification, either, but on how much more light the lens can catch compared to the naked eye. Without considering other conditions, here is a table that tells us what the maximum stellar magnitude is that we can reach in regards to the naked eye compared to the most widespread telescope diameters.

Telescope Diameter in mm	Δm (Increase in magnitude with respect to the naked eye)
80 mm (3.15 inches)	$\Delta m = 5$
100 mm (4 inches)	$\Delta m = 5.5$
150 mm (6 inches)	$\Delta m = 6.4$
200 mm (8 inches)	$\Delta m = 7$
250 mm (10 inches)	$\Delta m = 7.5$
300 mm (12 inches)	$\Delta m = 7.87 \approx 8$
2400 mm (94 inches) (*Hubble space telescope*)	$\Delta m = 14$
10,000mm (390 inches) (*Keck Telescope*)	$\Delta m = 17$

In short, therefore, magnification is just the way in which we take advantage of the potential that each instrument offers us, determined only by the diameter of the lens. From now on, never choose a telescope on the basis of how much it can magnify the image.

The various telescopes

Back in the day, telescopes were very simple instruments, made up of one lens for the objective and another smaller one that had the task of magnifying the image to make it visible to the eye. This lens, as with binoculars, is called eyepiece, or ocular.

During the course of the centuries various types of telescopes were born, all of them with an objective: showing the images in the sky as well as possible and maybe add some advantages such as lightness and ease of construction.

Now there are three big families of telescopes: the refractors, which use only lenses, the reflectors which employ only mirrors and the catadioptrics which, instead, unite lenses and mirrors. Inside these families, we have those that are called optical patterns, which means different ways of coupling the elements of the telescope.

Refractors are divided, for example, into achromatic, semi-apochromatic (or ED) and apochromatic, strange words that take into consideration the number and the quality of the lenses that make up the objective.

Achromatic refractors, the most economical, are made up of an objective formed by two lenses made with ordinary glass. Unfortunately, however, they don't restore perfect images unless they are very long, that is with a focal length at least ten times longer than the diameter. **The semi-apochromatic refractors** or ED always have two lenses but use special glass that restores much clearer and cleaner images. **The apochromatic** uses three, at times even four, to correct all defects and to give very clear images of the entire visual field. Unfortunately, they are the most expensive, the secret fantasy (and superfluous unless you're a photographer) of many buffs.

In the reflectors family we find the **Newtonian reflector** or Newton, the most economic and easiest telescope to build. This is formed by a concave mirror, shaped like a parabola, placed at the bottom of the tube. The primary mirror catches the light and concentrates it onto a small mirror placed in front of it, almost at the beginning of the tube. The job of the secondary mirror is to bring the light out and making it visible to the eye through an eyepiece. The Newtonian telescopes seem very strange because they're usually squat and the observation is done on the side, near the end of the tube, contrary to all the other types of telescopes.

Other reflectors are called **Cassegrain** and use distinctly-shaped mirrors to shorten the tube and render transporting the instrument much easier. To compensate, they cost more than their Newtonian colleagues.

Instead, the catadioptric use simple-shaped mirrors that produce intentionally defective images, to be corrected with what is called a corrective plate, which is nothing more than large eyeglasses placed at the beginning of the tube. There are different kinds but the more widespread variations are the **Maksutov** and the **Schmidt-Cassegrain**, both very compact and portable instruments. For example, a 200mm (8 inches) Newton telescope with a 2000mm focal length would be almost six and a half feet long and weigh at least forty-four pounds. A Schmidt-Cassegrain of equal diameter and focal is only 16 inches long and weighs 11 pounds!

The optical configurations most used by amateur astronomers. Above left, the classic refractor, the telescope par excellence. At right, the Newtonian reflector, the cheapest and strangest, because observations are made from the top of the tube. On the lower left, the Schmidt-Cassegrain, a catadioptric configuration available for diameters from 150 mm up (6 inches); and to the right, the Maksutov, a catadioptric configuration usually available for small diameters of 90-130mm (3.5-5.5 inches).

The mounts

The term telescope usually defines the entire instrument used to observe the sky. If we observe it well, we'll understand that each telescope is made up of at least two pieces: the actual tube, called optical tube, is the part that lets us observe and a support that lifts it off the ground, called the mount, often consisting of the classical tripod, and by a head which connects the support and the instrument.

When I bought my first telescope, I didn't even know what the term 'mount' meant; the merchant probably didn't know that well, either, since he simply sold me an instrument connected to a tripod. It seemed fairly solid, but when I began to observe the Moon at a magnification of more than 100, I understood that the appearance had fooled me, because that darned support vibrated every time I brushed against it. And it's better that we don't talk about what happened when I tried to follow the Moon, which, because of the Earth's orbit, moved quickly in the field. I had to go a little upwards and then a little to the right, manually moving the tube and being very careful to make gentle movements, otherwise I would have lost the angle. It wasn't exactly easy, especially because of the vibrations and the difficulty in making the necessary small movements. Yes, because if we observe at 100 magnifications, the support that holds up the telescope is 100 times more sensitive to movements and vibrations.

The lesson of this umpteenth disastrous experience was the following: a telescope requires an adequate support; otherwise, we'll get seasick while observing. This can't consist in a simple tripod like those used for cameras, but must be a real mount, solid and with sharp motion mechanisms to accurately follow celestial bodies.

My rickety tripod didn't have these mechanisms and, as time went by, its movements became harder and "jerky". My exasperation grew excessively inside me and exploded one evening while I was trying to point Venus without being able to frame it after a good five minutes; in a fit of anger, I kicked the tripod and caused the instrument to fall disastrously to the ground, cracking the internal lens of the objective.

And so, in the hopes of avoiding a similar situation, I beseech all of you: a telescope needs a solid mount to be used without going crazy!

The types of mounts

There are many types of telescopes, it's normal therefore that there are different types of mounts. Also in this case, however, the magical number is three, because this is how many families of mounts there are that are appropriate for our instruments.

Alt-azimuth (or altazimuth) mounts. This is the classic tripod, hopefully heavier and endowed with what are called micrometrical movements, handles that permit us to very gently move the telescope. The (alt-azimuth) mounts move along two directions: high-low (alt) and right-left (azimuth). It's simple to use and can hold any type of telescope except for Newtonian reflectors – for which there is another type of support that we'll see a little further ahead. However, this mount has a problem with projection: it was conceived to view the panorama, not the stars. Because of the Earth's rotation, the larger the magnification used, the faster the objects in the sky seem to move. With magnifications of over 50 times, you can see the movement with a telescope without difficulty. With magnifications over 200 times, the object will remain in the field for little more than a minute.

The alt-azimuth mount is nothing more than a heavy tripod endowed with, at times, micrometrical movements.

One could think, 'no problem; just point again'. And this is when we'll discover how inconvenient it is. While we won't be in the disastrous situation I described before with a dancing mount, where everything has to be done with your bare hands, we'll immediately realize that not all stars move perfectly back and forth or up and down. The rotation around the North Star takes place on an oblique circumference (we remember that, right?). Therefore, to follow the stars with the alt-azimuth mount, you have to move a little higher up and a little to the side, otherwise you'll risk losing the object. It's not that difficult, but at the beginning it can seem a little annoying. Maybe this is the reason that one day, long ago, someone had a great idea:

The equatorial mount. This is also attached to a normal tripod, but it has a strange shape and movements which at first seem annoying. Created only for observing the sky, it promises (a promise that is kept) to follow all objects by simply making a single movement. It's very convenient! All instruments can be put on or adapted to an equatorial mount, which becomes an even more convenient support if we buy a small and appropriate motor and make it do the movement for following the stars. In this case, the Earth's rotation is (annulled) and an object remains in the field even for three hours, regardless of how much it is magnified.

It seems too good to be true and in fact, there ae some contraindications that should be known. I have a personal experience for this circumstance, too.

My second telescope finally had an equatorial mount. When I reached home, I tried to set it up without even looking at the instructions. Strangely enough, I managed it; but, I didn't know how that strange, completely distorted thing called a mount worked. In fact, the telescope was all bent and if I wanted to go up or down, I couldn't do it.

That darned mount moved diagonally and, at times, when I tried to focus on a zone, it seemed as though it just didn't want to go there! Then I discovered that, at the base, near where it was attached to the tripod, there were other movements that finally let me move the telescope up and down and left to right. I had found a semblance of normality, but I was using the equatorial mount like a photographic tripod that wouldn't have given me any advantage over the stars. I understood it that same evening, when I couldn't observe anything; I decided then that it might be better to read the instructions.

The equatorial mount is a high precision system that permits us to easily follow the stars. It is mounted on a robust tripod and is very different from all those we have seen up till now.

It's hard to believe, but they were very useful (how strange, right?). I understood that for the equatorial mount to work well, it has to be regulated, even better, stationed, a simple operation that we will see in a while. There are also weights to insert for doing what is called balancing, something that has never existed with classic tripods. For strange movements, be patient: if we want to follow the stars we have to hold on to them and not think that it's working the wrong way.

Actually, the price is one of the biggest limits of the equatorial mount. If we have to set up an instrument that weighs more than 11-13 pounds, we need a very robust support that costs more than the telescope itself; that is why, if we don't want to take photographs and we make do, the alt-azimuth is an excellent choice; it also weighs less.

When low-priced Chinese productions didn't exist yet, this problem weighed heavily, because a mount that was appropriate for a small telescope could require a full month's wages. And then, without any more money, what would we have put on it? I imagine that it was with a spirit like this that a young American enthusiast, even poorer than me, decided to not give up. "It's better to observe and give up a little convenience than to have a nice mount on which I can only sit down," he'll have thought a little angrily. And so, with a bit of discarded materials collected here and there, he had a genial idea. That man was John Dobson, the inventor of the:

Dobson Mount. It's an extremely economical support that stands on the ground and, therefore, without a tripod (which isn't cheap), it can only hold the Newtonian type telescopes. Generally made of wood, it's a base that moves like the alt-azimuth on which you can set telescopes with a generous diameter. Therefore, Dobson telescopes are all those Newtonian reflectors used on a Dobson mount and represent the most economical instruments that are on the market. Sacrificing the entire complicated mechanical part (an equatorial mount is very complex!) you can concentrate energy and money only on what counts for a good observation of the things in the sky: the diameter of the tube. The Dobson mounts are therefore very simple instruments, often made of wood (at one time they were even made from cardboard!), completely manual and represent our last hope for observing without emptying our wallet.

Naturally, they aren't free of defects; they should be brought out in the open. Their movements are those of a tripod, so two movements must be combined to follow the stars, there are no motors that do it automatically for us (at least, not cheap ones!), so you can't take photographs except for a few of the Moon and bright planets, because these require long exposure and therefore, an

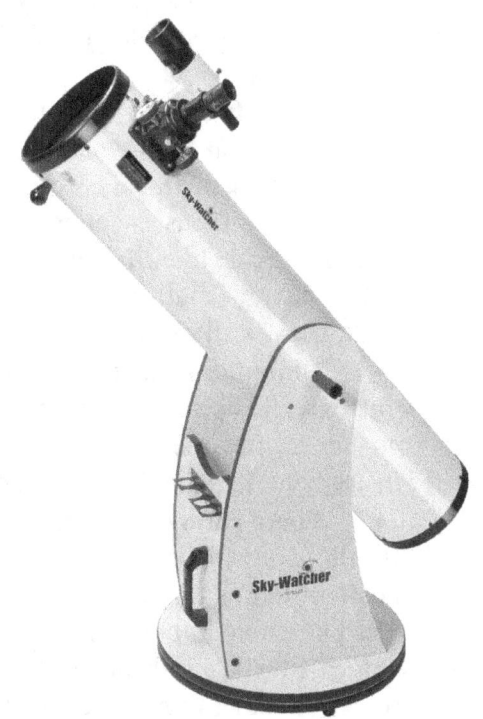

The Dobson mount is a very simple alt-azimuth and is paired with Newton telescopes with an ample diameter. The Dobson telescopes are specialized in deep-sky observations.

equatorial mount with a motor (motorized). They aren't aesthetically pleasing, but they love to be patched up with masking tape and glue when they break (not like the other mounts that require a trip to be repaired!). In short, the Dobson telescope is a sort of philosophy of life, so much so that many enthusiasts, especially on cloudy evenings, lose time squabbling over which is the best solution: an equatorial mount that does everything by itself, with gears made from very fine coal dust – previously digested by a yeti – or a wooden base that sometimes needs a well-placed kick to make it work but can be made at home with a few boards and a couple of tools. We don't like wars and the only thing we want to do is observe. If we're not filthy rich we'll have to take into consideration the instruments cost and choose the telescope on the basis of our personal needs, because the best is the one that we'll use the most and with the greatest constancy. Everything else is just gossip in the wind.

A few numbers on telescopes

Don't be frightened by the paragraph title. I won't present strange and complicated formulae that aren't useful for anything, but a few sizes that could be useful, at least to clear things up a little for our slightly (probably) confused minds. So then, let's see in syntheses a few sizes associated with the telescopes, trying to clear things up a little more.

- **Focal length (F):** It's the distance at which the telescope's lens is able to create the image of what it has framed. In simple telescopes, like Newton's and refractors, the focal is equal to the distance between the lens and the point in which the focus is found, right near the eyepiece. In the more complex configurations, like Maksutov and Schmidt-Cassegrain, the particular shapes of the mirrors lengthen the focal, keeping the tube short. It seems like a magic trick, but it's all physics,
- **Diameter (D):** It's the width of the lenses of a refractor or of the main mirror, called primary mirror, of a reflector or catadioptric telescope. The diameter is often measured in millimeters and is found on the tube itself of the instrument, together with the focal; and at another size:
- **Focal ratio (f):** this is a number that tells us how many times the focal length is larger than the instrument's diameter. But, the photographic luminosity – that is, the instrument's ability to take long exposure photography, just like photographic lenses – is also associated with the focal ratio. Small values, like f4-5 indicate a very luminous instrument; values higher than f10 refer to a rather dim telescope. This concept is applied only in photographing objects in deep sky, but not when observing them. Therefore, with an equal diameter, an instrument opened at f4 and one closed at f10 return the same luminosity to the eye. For those of us who only wish to observe, the focal ratio is of only marginal importance. It gives us indications in the achromatic refractories on the quality of the images because the focal ratio must be greater than f8 for them to be acceptable. At one time, a very low focal ratio – around f3-4 – in the Newtonian reflectors also indicated a worse quality of the optics because working such open mirrors is very difficult; but today, the differences in quality between an f4 instrument and one of f8 are imperceptible for our purposes.

Choosing the telescope

If we already have a telescope, then this paragraph is useless, at least until we decide to buy another. But beware: before changing your instrument, make sure you have taken complete advantage of the one you already have; otherwise, it would only be a waste of money. It's best to not waste money in these difficult times!

Which is the best telescope for us, to begin with and to also have great satisfaction? Mainly, it's the one with the largest possible diameter because it is certainly the most powerful. But beyond the problem of prohibitive prices, a telescope with a greater diameter is also heavier, difficult to handle and bulky. In short, it would be like getting your driver's license and trying to drive a Formula 1 race car: an idiocy!

The magic word in these cases is: compromise. In fact, we have to find an equilibrium between power, usability and, naturally, cost.

The latter is probably the most painful. Yes, because a telescope is a very high precision instrument that requires a lot of work to be built (mirrors and lenses must be worked to a precision 10 times finer than the thickness of a strand of hair!), and therefore it has a high price tag.

So, before analyzing which telescopes are the ones for us, let's ask ourselves a question: how much money should we spend for an instrument that lasts for a lifetime and which represents the door to a fantastic Universe? As accustomed as we are to technological objects that we could do quite well without, many often consider astronomy as a hobby that they can't afford. If anything, it's the opposite, because astronomy is a passion that stimulates the mind and body and certainly helps us grow better than a stylish cellphone. Let's remember that and remind our parents, too, if they are paying for the telescope. The nicest gift that they can give us and we can give ourselves is thinking of buying an IPhone, putting the necessary money aside and, instead, spending it on our first telescope which, at this point, will have a discreet power thanks to the price of a telephone that would no longer be "stylish" after a year. Just think that telescopes have been "stylish" for more than 400 years and that we don't foresee their fall for at least 100 years from now!

A telescope hardly requires any maintenance and if well-kept can last for more than 50 years. Are fifty years of the wonders of the Universe worth at least the money for a cellphone? They are probably worth even 100 times the amount, but I, naturally, am a bit biased.

Having said this, at present, you can actually purchase a small telescope even with less than $100. I don't advise spending such a low amount because what we'll have will be proportional to the price. A budget of $300-500 is perfect for our first instrument and maybe some accessories, which never hurts. If you look carefully you can spend even less than $200, but keep an eye on what you buy!

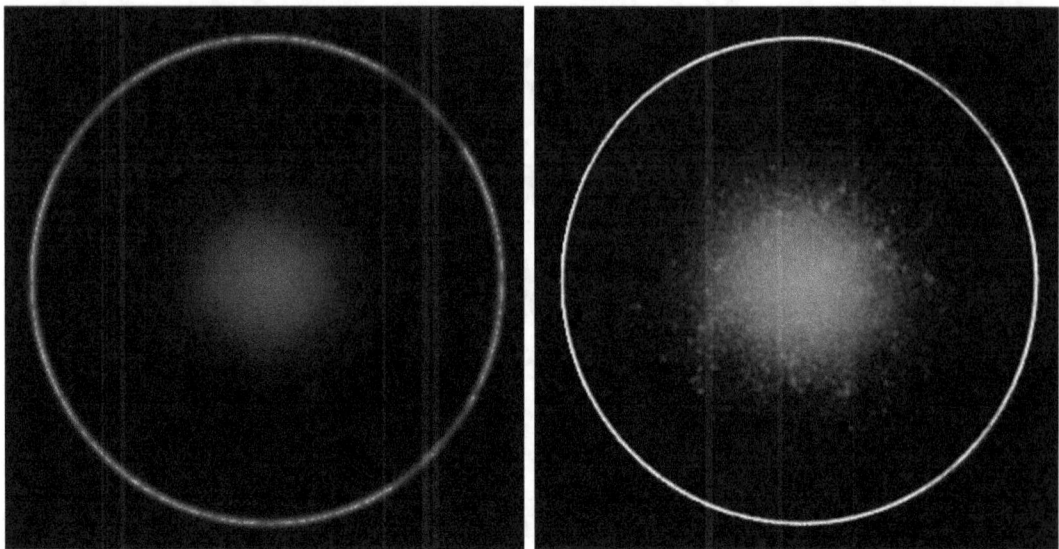

The diameter always wins. The greater the width of the lens, the better you can see dim objects and tiny details. In this example, on the left is the globular cluster M13 as it appears as seen through a small 80-100mm (3-4 inches) refractor and, on the right, as observed through a reflector with a 200mm (8 inches) diameter. The difference is enormous. The small diameters of 80-100 mm (3-4 inches), in this case with lenses, work well with planets, even though a greater diameter always wins, even in these cases.

Which telescope is right for us?

It's the question of all questions. Refractor or reflector? Equatorial mount to follow the stars or else the alt-azimuth or even the Dobson? Everything depends on what we want to do and how much space we have.

If we like to explore the Universe with our own efforts, we don't have problems with space, we're not interested in taking pictures except maybe some shots of the Moon and the brighter planets and above all we have a dark sky at our disposition, then the magic word is diameter. The best choice is to concentrate on a Dobson telescope. With little more than $300 you can buy a Dobson with an aperture of 200mm (8 inches) and with less than $500 you can take home a 250mm (10 inches) telescope. These are perfectly respectable diameters that will let us see thousands of objects amazingly well. The globular clusters will teem with subtle stars, the planets will be beautifully colored and full of details, it will seem like you are walking on the Moon, the nebulae will show intricate patterns and we'll be able to glimpse hundreds of faint galaxies.

However, managing a 250mm (10 inches) diameter telescope isn't easy, especially if we are young because it weighs a lot and there's a risk that it is taller than we are. A good compromise is to get the smaller brother of 200mm (8 inches).

The Newtonian telescope on a Dobson mount is the most powerful we can allow ourselves on a fixed budget. If we want other conveniences we'll have to either get a smaller diameter or increase the money to be paid out.

If we don't have a lot of space or we prefer a more manageable instrument, a Maksutov or a Schmidt-Cassegrain are the ones for us, but we'll need an equatorial mount. The instruments allow us to take a few more photos than the Dobson, but we can forget about the beautiful shots that we see online. A much more complicated instrumentation is needed for those (among which a much sturdier equatorial mount) which is difficult to find lower than $1500 (optimistically). With our $500 and a very compact tube, we can arrive at a complete telescope of about 127-130mm (3 inches).

The question of available space is very important. When the 250mm (10 inches) Newton arrived at my house, I could hardly stand still and I put them together right away on the balcony – no wider than 39 inches – of my new house in the city. The result? In order to observe I had to lean over the balcony to reach that darned eyepiece hanging into the void at more than 39 feet above the ground. Thank goodness my mother never saw me or else she would have fainted!

It couldn't be helped – I had to take it to my grandparents' house where there was more space; but, once again, there was another problem: it wouldn't fit in my mother's SUV!

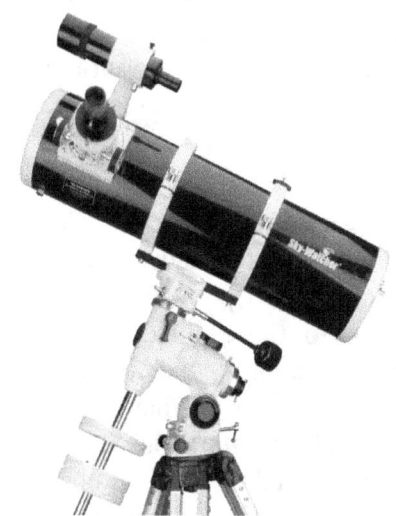

Newtonian Telescope, 150mm (6 inches) on an equatorial mount: an excellent instrument for beginners, without completely precluding the possibility of taking photographs.

Fortunately, my father had a construction company and he took it for me in one of his vans, otherwise I would have had to call a moving company to do it for me!

The sky is also very important. If we're in the city and can't move far from the lights, there's not much sense in getting an instrument useful for deep sky observations like a Dobson, because binoculars will let you see better. Only planets and double stars can be observed satisfactorily in the city and this is why an achromatic refractor with at least an 80mm (3 inches) diameter and with a focal length of at least 800mm (30 inches) – the minimum necessary so those colored rings around the objects won't be too annoying. I was highly satisfied observing the planets with my second telescope, a 90 mm (3.5 inches) refractor with a 910mm (35 inches) focal length. An instrument of this type can give the same details as a telescope with 114-130 mm (4.5-5 inches) mirrors, from the Moon and planets, and is easier to manage. I advise against getting achromatic refractors with shorter focals, because the defects in the images start becoming annoying. The chromatism in particular – the technical name for the defect that shows the colored rings round the objects – affects the resolution we could reach, in addition to disturbing our view more than just a little. And the apochromatic? Well, they would be perfect, but I could never afford them and I would probably never get them if I only wanted to observe the sky.

If we have a higher budget available and a good, strong passion, we could think about investing around a $1000 in the classic telescope par excellence: the legendary C8, a Schmidt-Cassegrain with a 200mm (8 inches) diameter, 16 inches in length and a weight of only a few pounds; it's easy to carry and as optically powerful as a cumbersome Newton with an equal diameter. This telescope, also because of its excellent optical quality, has gained fame in the field as a factotum instrument that can give satisfaction for many years. Up until the early 2000s, its purchase was prohibitive because together with its mount it cost twice as much as a worker's monthly wage, whereas now, in some cases, it can be found for even less than $1000, all inclusive.

What about automatic pointing?

During the last decade telescopes with complex (and fragile) electronic mounts, promising to self-point at objects, have been popping up like toadstools. They are instruments for those who don't have much time or are too lazy, because they often avoid having to search for the object. A few pushed buttons on the hand controller that controls the mount and the instrument is directed toward the object. At this point, the spirit of amateur astronomy is completely lost: that of trying to discover through our own efforts. In fact, one continues to be a passive slave of a computer that tries to think for us. You might as well keep surfing the web to look for photos that are taken by other enthusiasts and certainly nicer than what we see with our own eyes.

I understand that what I just said is my opinion and as such could be either right or wrong, so here are a few more realistic points to help understand whether we want a computerized telescope.

A small computerized achromatic refractor of 80mm diameter (3 inches). Is it better to spend $400 for an 80mm electronic telescope or $300 for a 200mm (8 inches) Dobson? Personally, I have no doubts at all: no matter how many steps ahead technology can go, an instrument's power is always in its diameter.

Automatic pointing or GOTO is a fragile and, above all, expensive gadget. With our maximum $500 budget, we can buy a 114mm (4.5 inches) Newton telescope or a small 80mm (3 inches) refractor, and it's really a lot of money for what we'd really be able to see.

Then, there's another thing known only by a few. There is no telescope anywhere in the world that does everything by itself. Don't believe those who would like to sell us an impossible dream (?); rather, let's accept the fact that looking for objects isn't difficult, if we want to learn, and it's perfectly within everyone's reach.

In fact, the GOTO must be calibrated every evening by pointing a few stars by hand. If we don't know the sky, how can we calibrate it? Not to forget that this procedure doesn't always work, making us lose precious time.

In a perfect world, where we had tons of money available, we could get a nice 250mm (10 inches) computerized Newton with a strong mount that, when necessary, could even prepare a midnight snack for us and pamper us with a nice massage. But if our resources are limited and the desire to explore the Universe with our own efforts is still very much alive, then the only thing that really matters is the instrument's diameter. Let our brain think about finding the objects. Let's do ourselves a favor: let's not allow our brain to fizzle out!

Which brands should we choose?

All inexpensive telescopes, therefore under $2-3,000, are produced in China by few factories which are then sold to merchants around the world with different names.

1) The brands to avoid. A name out of all that you should stay far away from: Seben. They are telescopes often sold on e-Bay and/or by incompetent people, flavored with very long descriptions that are full of inaccuracies that try to attract ingenuous dreamers. Often, they are even sold as professional telescopes. Well, professional telescopes are as big as a room, weigh as much as 10 cars and cost several million dollars. There are no professional telescopes sold commercially. If anyone wants to sell you a professional telescope for a few hundred dollars, it's better to turn around and run away without looking back. The Seben telescopes are very cheap and are therefore very tempting, but they are literally the bottoms of broken bottles. I always agree about not wasting money buying a billion dollar instrument, but spending too little is also a waste because that object will be found worthless;

2) The good and inexpensive brands. The telescopes that are found in specialized shops that bear the names Celestron, Meade, Bresser, GSO, Skywatcher, Geoptik, Orion, Ziel, Konus, TS and all those that propose similar prices, sell good quality telescopes at interesting prices and are the instruments for us;

3) The dream brands: Televue, Takahashi, Borg, Planewave, RpAstro, Astrophysics, TEC and so forth are the impossible dreams for all enthusiasts, because they produce telescopes of the highest quality; but even just thinking about them could empty our wallets, so for now they are outside all limits (and for me they always have been). The truth is that if we work around things a little, we don't need these super-telescopes to just observe the sky.

The telescope with which Galileo Galilei discovered Jupiter's satellites was of such bad quality that these days no one would be able to produce it even if they wanted to.

So, the good quality commercial brands, like the ones mentioned above in point number 2, are good for us. The ones in this category are all the same and the only choices to be made are a question of esthetics or a more advantageous offer than another.

From whom should we buy the telescope?

The telescope should absolutely be bought from specialized retailers who, therefore, have good quality instruments (no astronomy store would dream of selling a Seben, for example) and can also help in the choice of everything that is needed. It's best to not go to your trusted optician, as was done at one time. I went straight to the optician right downstairs in my same building for my first instrument and the owner knew how advise me fairly well. But 5 years later, when I had to choose another instrument, I noticed that the owners of these activities had by then forgotten all about astronomy. I knew more than they did and this, when a person is young and needs to make an important choice, just isn't good.

Shops specialized in astronomy are rare at this point and concentrated mainly in the big cities. If they're far from your home, you should send an e-mail, call them on the phone or else go in person. If we have a pretty good idea of what we want, we can even turn to some big retailers who usually have lower prices and are just as trustworthy.

90mm (3.55 inches) achromatic refractor with a long focal on an equatorial mount EQ2. The suggested instrument for observing planets, the Moon and double stars. Average price: $200.

130mm (5 inches) Newton on an equatorial mount EQ2. Suggested instrument for those with limited money who have no specific preferences in what they observe. Average price: $200-300.

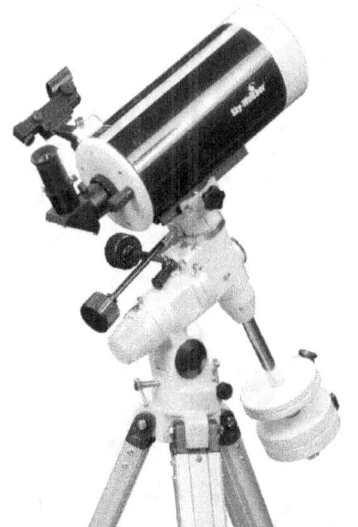
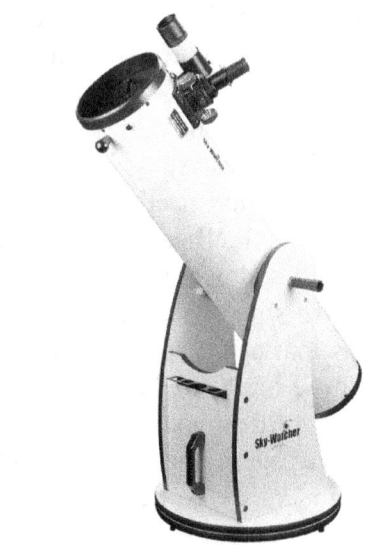

127mm (5 inches) Maksutov telescope on an equatorial mount E3.2, appropriate for those who mainly want to observe the planets, the Moon and double stars and have very little available space. Easily transportable instrument. Average price: $400.

200mm (8 inches) Newtonian telescope with a Dobson configuration, for the observers with no problems with space, who have a dark sky and want to mainly observe objects in deep sky. Average price: $300.

How do we make the telescope work?

We have chosen and bought the telescope – maybe by following my advice given above – or else, we already had one. Perhaps we discovered that we had an infamous Seben and, after having read the preceding pages, have been convinced to exchange it. Don't do it. I would never advise buying this type of telescope but if we already have it, it's better to take the maximum advantage. Since, just between us, your first instrument will be submitted to every kind of torture, a bit like your first car, having an instrument in your hands that can be literally torn apart just makes us act with more freedom, because we know that if we destroy it, we won't have lost that much.

With the instrument probably in front of us in this room, or already set up outside in some manner, we can hardly wait to have a nice observation and explore the Universe. I fully understand the feeling because my instrument, as I'm writing these lines, is outside waiting for me, already pointed at the first quarter of the Moon. Yes, I've already observed it thousands of times by now, but having a great passion implies never getting tired of repeating something that always seems to be the same over and over.

In fact, now that I think about it, I'm going to go take a look at the Moon and then I'll explain what operations you need to do to take on the magical world of telescopic observation. A preliminary suggestion: if the telescope is already outside, it's not by chance.

Here we are, back together again. Where were we? Right, how to make the telescope work.

If we're lucky, we bought it already set up or almost, but most likely it's still in the box and needs to be set. Don't worry, I'll tell you what to do. Put this book down with a bookmarker at this page and read the instructions that came with the instrument. These are your best allies in this phase. After you've finished, start reading again and maybe we'll begin to understand something more, together.

If we've already started reading here, we've probably been able to more or less put a large part of the pieces together, even if we don't know where some of them go (it's normal!) and we don't have any idea how to move the work of art we've built with a bit of effort. This is why,

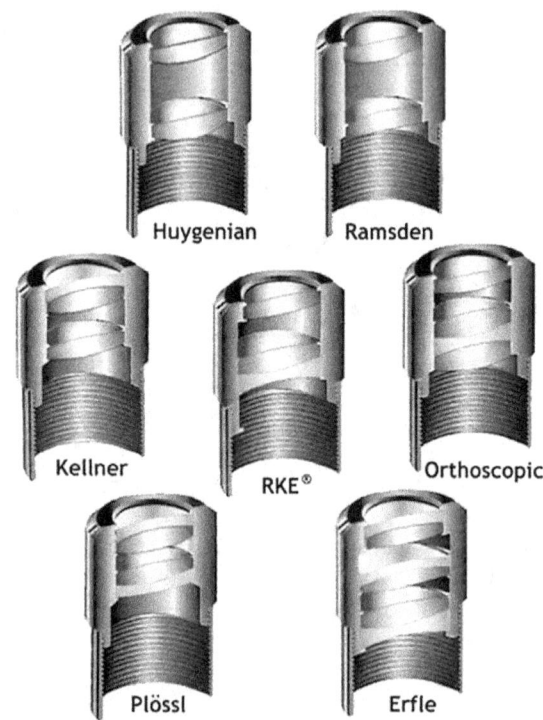

Optical schemes of the principal types of eyepieces. From the simple and qualitatively modest *Huygens*, to the complex and qualitatively excellent *Erfle*.

I hope, what I'm about to say will help to better understand a few things.

Well, first of all, let's start with the optic tube and from what we should do to see clear images, then we'll move on out to the field and gain confidence with the positioning, but only if we have an equatorial mount. If our telescope is on a Dobson or alt-azimuth mount, we're already ready to observe...

How exciting, we're almost there!

What do we need for observing?

A telescope is essentially a large photographic lens and without an indispensable accessory, it can't furnish images that our eye can see. This accessory is called 'ocular' or eyepiece and we saw it briefly when we were talking about binoculars. In telescopes, we can change eyepieces as we like and, fortunately, we can only insert one on at a time. In the refractors and catadioptric the eyepiece holder, or rather, the casing in which this precious optical accessory is inserted, is found at the bottom of the tube. It's found on the side of Newtonian reflectors.

At this point all telescopes use two standards for the diameter of the eyepiece barrel, which is that portion that embeds it into the casing. So, there are eyepieces with a 31.8mm (1.25 inches) barrel diameter and others with a 50.8mm (2 inches) diameter.

The latter are usually reserved for the more expensive and powerful telescopes than ours. The standard 31.8 (1.25) is more widespread. This means that we can use all, and that means all, eyepieces produced with a 31.8mm (1.25 inches) diameter for our telescope. There's no need to specifically buy one by the same brand as the instrument.

The eyepieces are found in commerce with diverse optics diagrams and prices. A good compromise is represented by the Ploss diagram which joins a good quality with an acceptable price. However, keep two things in mind:
1. The quality of the image that we'll observe is determined by the part with the lowest quality so if we have a good telescope but a terrible eyepiece, we'll always see badly;
2. Telescopes change but eyepieces always stay the same because they can be used on any instrument; therefore, we can make a bigger investment when the time comes.

Anyway, we don't to buy eyepieces for now because we already have one or two that came with the telescope.

There is a number written on all eyepieces, on the side where the group of lenses is held together. That is the focal of the eyepiece, expressed in millimeters and it's very important. Why? Because it gives us indications on the magnification we'll have using it with our telescope.

The **magnification** of the image is given, in fact, by a simple formula:
The telescope's focal *divided* by the eyepiece focal.

The telescope's focal length is found on the instrument, usually on the tube near the eyepiece holder. So, an instrument with a focal length of 1,000mm used with a 20mm eyepiece will give a magnification equal to $1000/20 = 50X$ (it works fine also with inches!). Simple, right? The appropriate eyepieces for observing objects in deep sky should give magnifications between 30 and 50 times, while for planets it would be useful to have a magnification of at least 150-200, according to the telescope's diameter.

Focusing

In many telescopes, the eyepiece casing can change position thanks to two handles that are found on the same mechanical group. This is the mechanism for focusing. Turning them we can focus on an image from a few dozen yards away all the way to the most remote regions of the Universe.

In the catadioptric instruments, the focus consists in a small handle detached from the eyepiece casing, on the breech of the instrument (the rear part, where you look from).

The possibility of regulating the focus in all telescopes is a wonderful invention because if we have sight problems, we can observe without glasses, unless the defect is astigmatism. If we are near-sighted or far-sighted the images will be perfectly focused even without glasses or contact lenses.

Aligning the finder

We'll have certainly noticed that mounted parallel to the instrument there is a small, curious looking scope or at least there should be. This accessory is called finder and will be our best friend during the evenings when we observe.

First, though, we have to gain its trust by carrying out a very important operation: an alignment. This small scope, which will magnify from 6 to 8 times, is necessary, in fact, for pointing the celestial bodies to observe.

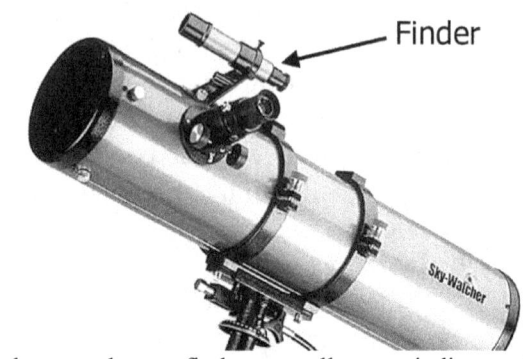

All telescopes have a finder, a small scope, indispensable for finding heavenly objects.

The telescope magnifies too much and finding even the Moon can become extremely complicated. The solution is obtained by using the finder which, with a very small magnification, will easily allow us to find the object.

Let's take the telescope outside during the day and, without being too shy, let's approach this interesting viewer. We can see some crossed wires that will help us perfectly center on the object.

The only thing required is that the lookout points exactly what the telescope sees and we have to do this every time we dismantle it from its base or transfer the instrument by car. The alignment of the finder should be done by day because it's easier that way.

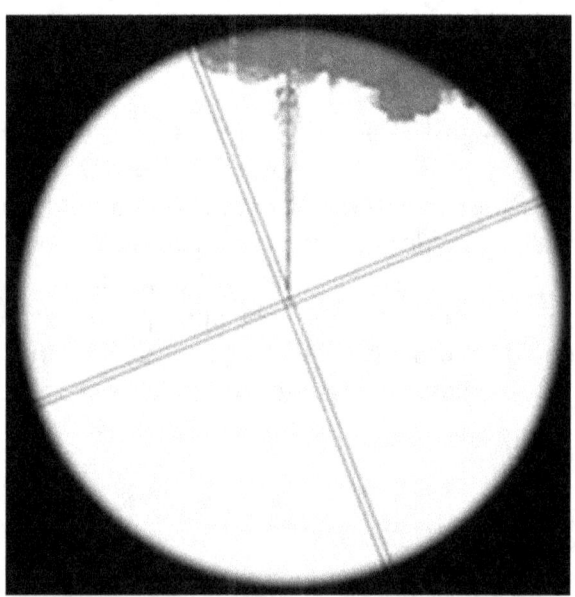

When an earthly detail centered in the ocular is also at the center of the crossing of the lookout, the latter is aligned and can be used to point heavenly objects.

With the telescope set up on its mount, we insert the eyepiece that gives us the lowest magnification (the one with the longer focal!) and point a detail that's at least a hundred yards away: a street light, the belfry of a church, a TV antenna, a rooftop...

Let's tighten the screws of the mount and now let's look at what's inside the finder: we surely won't see the central detail that is in the center of the telescope's field of vision. So then, we have to move it, but how? There are at least three screws on the support that can be rotated: these are what change the finder's orientation. Let's move them in different ways until the center of the crossing frames the exact center of the field present in the eyepiece. Let's do it so that the screws are tight – otherwise we'll immediately lose the alignment – and it's done. Now, we'll point every object that we'll want to observe with this precious friend, first.

Some telescopes with a small diameter don't have an actual scope as a finder but rather a sort of viewer with a red dot in the middle. This is usually already aligned, but on the chance it isn't, the procedure to carry out is the same one we have just seen.

Upside-down images

Just as we were trying to align the lookout we realized that the images we saw were upside-own. How is that possible? Is there something wrong? And yet, they were right with the binoculars!

Never fear, everything is fine. All optical instruments furnish upside-down images. This effect is corrected with the appropriate prisms in binoculars. It's left alone in telescopes since the concept of below or above loses its meaning in space.

However, if we want to use the telescope for on-ground observations, there are two accessories out there for straightening out the image: one is called mirror diagonal and the other, rectifier prism.

The diagonal doesn't make us see upside-down, but it inverts right and left because it uses a simple mirror, whereas the rectifier corrects everything because it employs prisms as with binoculars.

The mirror diagonal is useful with posterior-focus telescopes for conveniently observing objects that are very high on the horizon.

I don't advise using either one of these two objects for astronomic observations if avoidable because, among other things, they can't be used on Newtonian telescopes. The mirror diagonal can be useful only on telescopes where observations are made from the base of the tube because, inserted in the eyepiece casing before the ocular, it allows us to be slightly more comfortable while observing, especially with objects that are very high above the horizon.

I was without one for a long time and when I had to observe celestial bodies almost at the zenith with my Schmidt-Cassegrain, I had to literally lie down on the ground and crawl below the ocular: a decidedly uncomfortable position!

The movements of the equatorial mount

A large part of the equatorial mounts found out there (maybe all of them, now) are called German equatorial mounts and are made like this:

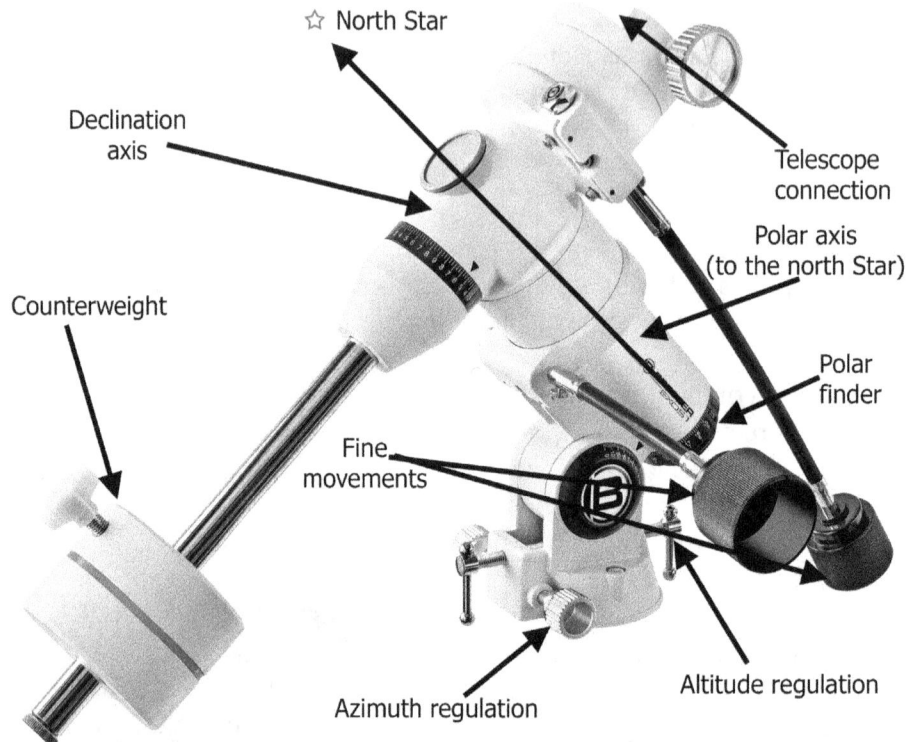

Typical German equatorial mount that equips amateur telescopes.

The supporting structure is developed around an axis that is inclined in regards to the tripod. This is called the polar axis and it's the one that must point the celestial North Pole which is identified approximately with the Northern Star.

The movements at the base of the mount, which make it move completely as though it were a normal tripod, are there exactly for this reason. In particular, we can regulate the height and what is called the azimuth, nothing more than orientation in regards to our horizon. This must be acted upon at the beginning of an observation, during the alignment phase; after that, they shouldn't be touched again because they are used only for a correct orientation, not for pointing and following objects.

It already seems terribly complicated, but it's one of those cases where practice is much easier than the theoretical explanation, so let's not create too many problems; let's wait for a calm evening and bring out the telescope so we can understand out in the field how to position this danged mount. If we are the happy owners of a Dobson or a refractor on an alt-azimuth, we can skip this step, take an evening off to do what we want and set up an appointment after the next few pages.

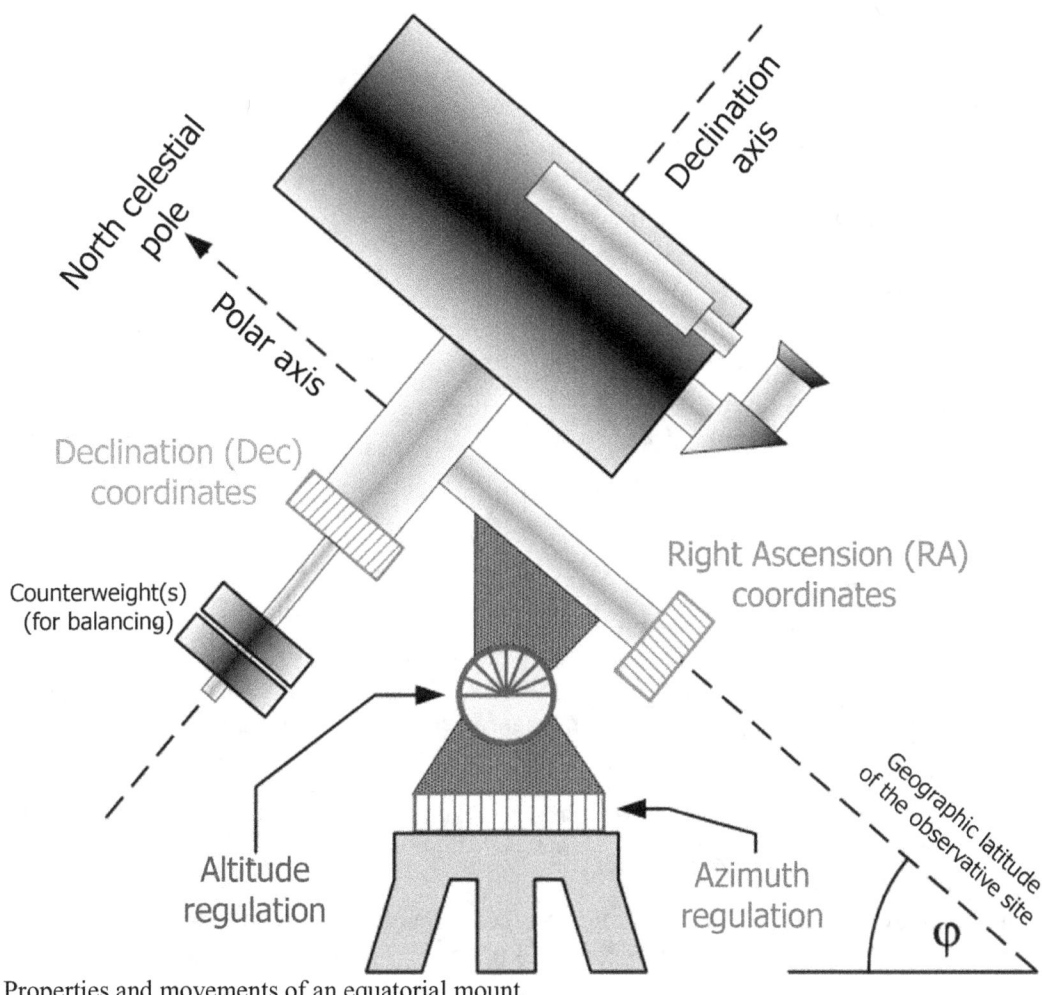

Properties and movements of an equatorial mount.

Aligning the equatorial mount to the pole

The positioning must be done at night, with the North Star visible because, in this case, the precision of a compass isn't enough, and without the telescope being set up.

If our observation site doesn't allow us to see the North Star we can change spots, or else be guided approximately by the compass, keeping in mind that the magnetic North Pole indicated by the needle is about 5° further west than the real one.

If we see the North Star, everything will be easier and more precise.

First of all, it's important that the base of the mount is more or less flat. If we are on a slope, we'll have to work on the length of the tripod's legs. Levelling it, as it's called, is a very useful operation for avoiding some unpleasant complications, so it's better to lose a few minutes now, perhaps even with the help of a small level, rather than spend the whole evening swearing. And since we're at it, before putting the level on the tripod, let's rotate it so that the polar axis of the mount already points sort-of toward the North Star.

After finishing levelling, and working on the movements at the base now, we have to tilt the axis to an angle equal to our latitude. A graduated scale will tell us at what point we are. The next move consists in working on the other movement to change the orientation and to point precisely at the North Star.

The more complex mounts have a small finder embedded in the polar axis which helps give an even more precise alignment. Actually, the precision reached by just looking is more than sufficient for observing, but if we have an equatorial mount with automatic pointing, it's better to refine the alignment; otherwise, it won't find objects for us. How is it done? It's simple: just take a look in the finder. Inside, there are drawings engraved, generally Cassiopeia and the Big Dipper, which

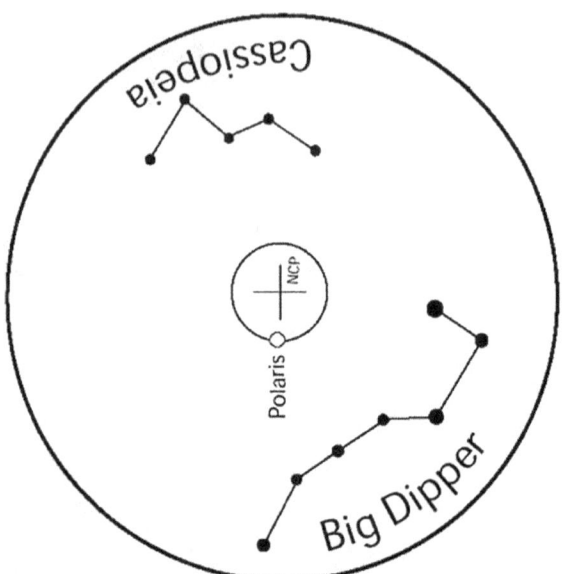

The most advanced models of equatorial mounts have a little finder embedded in the carrier axis, and this is what we see inside it. To carry out a precise alignment, the right ascension axis must be rotated until the figures of the constellations fit together; then, work on the movements at the base (height and azimuth) again, in order to put the North Star into the small circle near the central area.

will be our references for understanding where the celestial North Pole is found. We have to rotate one of the mount's axes (the ones that move the telescope, not those at the base!) until we find the correct orientation of the constellations, then we'll block their movements. This is necessary because the North Star isn't right on the celestial North Pole, and this way we ensure that we point it. Now, using the regulation movement, we have to make sure that we bring the North Star into the small circle marked in the finder. Once we've done so, we'll block the screws and we won't touch these movements again.

We're ready to put on an important part of the telescope: the weights for balancing it.

Balancing the telescope

Because of the strange movements that equatorial mounts have to make, we have to somehow balance the weight of the telescope; otherwise, at certain points of the sky it could even tip over.

The equatorial mounts sold in series together with telescopes already have the weights to make the balance. Here is how to do it safely.

First, let's set up the telescope, being very careful that the cradle that will be hosting it is pointed toward the North Star. In fact, this is the mount's most stable position when it isn't balanced and therefore, we'll keep it from tipping over disastrously. Let's be very careful that the screws of the axis are tightly blocked so that the tube doesn't slide.

Now, let's gently take the weights (or the weight) and insert them (it) into the special metal bar found on the part opposite the telescope. It's impossible to not recognize it.

The order of the operations can also be inverted: some prefer putting on the weights first and then the telescope. It doesn't matter, just be careful.

We tighten the weight's screws well, gently loosen the right ascension axis and rotate the telescope until we bring the bar for the counterweights parallel to the ground. Our task now is to move the weight and, if necessary, add another to make sure that the telescope – with the axis being free to move – is balanced, like a spoon on the tip of a finger. It's not a difficult operation, and with this, the telescope is almost ready.

If we really want to be fussy, we have to check the tube's balance also on the declination axis. Still in this position, we'll block the axis we've just balanced and loosen the other. If the telescope turns upward or downward, we should balance it by moving the spot where it is hooked up to the mount ether further forward or backward. The balancing operation, besides rendering movements easier and smoother, allows possible searching motors to work without being pushed too much and thus guaranteeing years of searching the skies.

We're truly ready now, but maybe we have to wait a little while longer, because I forgot to tell you a couple of important things.

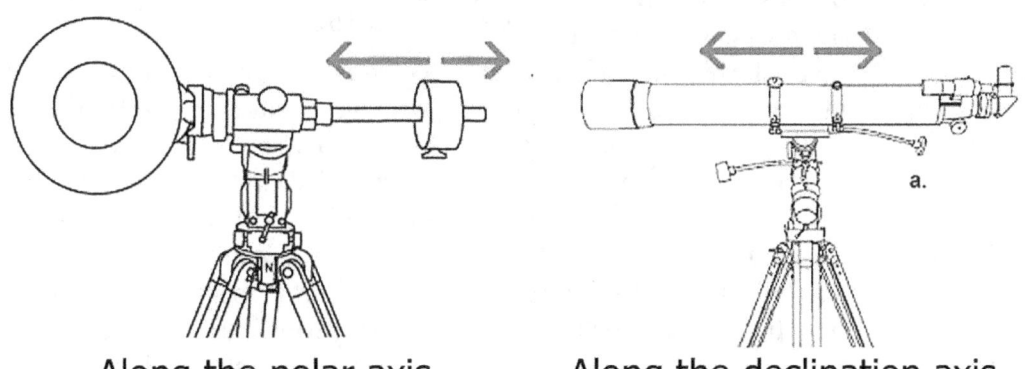

Along the polar axis   Along the declination axis

Balancing the equatorial mount is a simple but important operation.

The Acclimatization

To give its best, the optics of a telescope should have the same temperature as the surrounding environment. When we carry the telescope from a hot environment to a colder one, especially in winter, the images will initially seem to be much more jumpy and blurry than usual. So, the acclimatization is a phase that would be better to do if we decide to observe with high magnification. How is it done? Simple: take the instrument out at least half an hour before beginning the observations. Is that all? Yes; at times things are as simple as they seem...thank goodness!

How to set up the GOTO?

If our equatorial mount doesn't have automatic pointing, we're ready to begin. In the contrary case, we should decide whether or not to begin the alignment (or calibration) procedure of the computer. This helps only and exclusively if we want the instrument to find the objects in our place. If we want to do it alone to learn, or because we have to point bright objects, we can skip the procedure All the computerized equatorial mounts can follow the objects without having calibrated the GOTO. This is an important point to understand because I've often met amateur astronomers who, only to point the Moon or Jupiter, have spent ten minutes or more carrying out this boring phase. Meanwhile, I, with my equatorial mount and sidereal motor, observed a few seconds after I had mounted and aligned it.

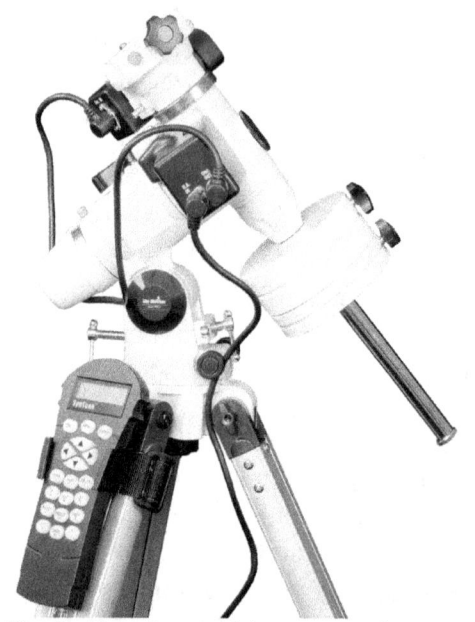

The SynScan is one of the most used automatic finders and equips the Chinese mounts of the EQ series produced by Skywatcher.

You don't need to do the alignment every time, because every motorized mount follows the objects anyway once we have found them with our hands and locked the axes.

On some models, like the white-colored mounts of the EQ series (of the Skywatcher brand), we have to activate the sidereal tracking (as it is called) in the keyboard menu, but this operation is much quicker and simpler than aligning. Instead, if we have decided that we need it (and I won't deny that sometimes it's convenient!) then the mount's manual will be our best ally for understanding it without becoming too angry.

A little advice, though: don't let electronic and computer problems ruin our desire for the stars. If, some evening, the GOTO drives us crazy, let's remember a gift that never goes out of whack: our intelligence. We're able to find the objects with our hands!

The Big Moment: let's observe with the telescope

Good, after all this theory on how a telescope works and which regulations have to be done, it's time to enjoy the first evening as full-fledged astronomy buffs. It's an important step, our sky baptism, and we don't want to arrive unprepared, of course. And this is why it's better to plan it so it will become a success.

First of all, it's better to have done a little practice with the mount and know how to align the finder. Let's study the eyepieces, we know how many magnifications they give us, how that strange equatorial mount moves… in short, let's try to simulate an evening now while there's still light, because it will be a little more difficult to do at night.

Then, we have to understand what we'll be observing. Because this time, contrary to binoculars, it will be very difficult to run freely around the sky and be explorers. Of course, we could bring along our binoculars (in fact, we will bring them!), look for objects and then try to point them with the telescope, but for this first evening we can settle with heading toward a few luminous stars. In fact, why don't we choose an evening with the Moon and begin with precisely that? Be careful that the Moon isn't in the full phase or very close because otherwise we won't see any details. The Moon can be observed very well in near the first and last quarters (around 3 days before and 3 days after—always keep this in mind!).

Therefore, we don't need to be under a very dark sky as long as we don't have street lamps beaming down at us. I already know, though, that after pointing the Moon, we'll want to see something else. True, but what? A good place to start could be the objects we tried to observe with binoculars a few evenings ago. We already have the star-hopping maps at hand, and we'll do that with the telescope's finder.

It's best to not do too much on this first evening, because we've already done a lot and our main objective is to gain confidence with the new instrument and understand how it works. Everything that comes after will be earned.

Let's prepare all our tools ahead of time. If we have to go by car, it's best to do so when there is still light from the sunset and set everything up when we can still see a little.

The first thing to do, immediately, is to control the finder's alignment: without that, we can't do anything. If you can still see a little bit of the panorama or if there are lights on the horizon, let's point those with the finder and see if they are in the center of the field framed by the telescope. If not, we have to repeat the alignment operation.

If we have an equatorial mount, let's remember to do all the phases we've seen just as soon as the North Star is visible.

Let's not be in too much of a hurry or anxious; let's stay calm. There will be a lot of delicate moments, especially this evening, but nothing will happen if we make a mistake or if our attempts aren't successful right away.

With the telescope set up and the finder aligned, we're ready for our very first observation. Destination Moon!

Let's insert the eyepiece with the longest focal into its holder; then, let's direct the instrument toward our satellite.

Remember that if we have an equatorial mount we'll only have to touch the levers which unblock the axes and make the instrument move a bit strangely. Never move the levers at the base because they only serve for aligning!

Just a few seconds for finding the Moon in the finder and understanding where to move with the upside-down images; then, when it is centered, we'll block the axes and savor our triumphal moment. Let's get our eye near the eyepiece, focus and let ourselves be carried away by the infinite beauty of what we are looking at... absolutely amazing; I can feel it even while writing these lines.

The Moon is always enchanting, with any instrument. A dreamy vision of a world that is so close for the Universe, yet unreachable for us human beings.

Let's observe, just let ourselves go, lose ourselves in the beauty of our natural satellite full of holes while at the same let's dare, let's experiment, let's try. Let's attempt to understand where the most opposed details are. Yes, they're all near the border, right there where the dark and the light meet. This region is called terminator and it's exactly here that we'll see the most incredible details.

Let's try to increase the magnifications by changing the eyepiece; let's explore that desolate place almost 239,228 miles away, where 6 astronauts walked a long time ago. The Moon now appears as it really is: another world, another land, another panorama hung up there in space as though by magic... We are officially amateur astronomers but there is still a very long way to go.

For now, though, I have nothing left to say. We're free to do what we want and we're perfectly entitled to be alone in this very unique moment.

Some advice for better observations

The first evening out with the telescope is over, but the desire for the stars isn't; at least, I hope not. I can't know the details, but I can think back to what I went through and still savor that unique bittersweet taste. Sweet, because the Moon, in the end, excites us; bitter, because maybe we thought that observations with a telescope were a little easier. No problem; with practice this aspect will disappear, little by little, and all that will remain will be the joy of being able to observe the Universe directly with our instrument, without having to see it on the computer screen or on the pages of a book.

Now that we're home and a little more relaxed, we can try to understand what went well and what went wrong and to clear up some doubts on how to see more and better.

I've already said everything about the part dedicated to aligning the mount and pointing the objects, but I'll repeat it again: you need to practice – there's no way around it. When pointing the Moon, you'll surely have tried to observe it at the highest possible magnification and will have noticed a couple of things: the image was darker, perhaps a bit blurry and seemed to boil, as though it were immersed in a pan full of water.

We're dealing with two aspects that must always be considered, the first of which we already saw when we talked about the power of a telescope.

We've discovered what happens if we magnify an image beyond its optimal value, or rather, beyond what I have called the maximum useful magnification. The magnification is usually increased for observing very bright objects that present very few details, so we can take advantage of its resolving power.

This power is given by the diameter of the telescope according to a simple formula:

Resolving power = 115 / telescope diameter

This formula, invented by an astronomer named Dawes, tells us that a telescope with a 115 millimeter diameter has a maximum resolving power of 115/115 = 1 arc second.

Oh my, what is this strange unit of measure? There's nothing simpler. The degrees with which we measure the distances and the dimensions of the celestial sphere are large angles for certain objects, so a more appropriate unit of measure is used. In fact, a degree is composed by 60 arc minutes (symbol ') and an arc minute is composed of 60 arc seconds (symbol "), just like an hour in the clock is formed by 60 minutes and a minute by 60 seconds. Therefore, an arc second is a tiny unit of measure and represents a three thousand six hundredth part of a degree. We finally solved the mystery that arose since page 82!

Coming back to our telescope, if the resolving power is 1", it means that it can let us see details up to these dimensions. As seen, this is a potential reached by magnifying the image up to a maximum of 2.5 times the diameter of the objective expressed in millimeters. If we go further, it's like continuing to endlessly enlarge an image on the computer: the resolution doesn't change, but seeing it larger it seems blurry and darker.

However, there is still something that we don't know and it regards that boiling of the lunar image that seems truly annoying at times. What caused it? Is the Moon immersed in water? Unfortunately no (it would have certainly been fascinating!). With our own efforts, we have discovered the number one enemy of high resolution observations: **atmospheric turbulence.**

The air above our head seems calm and transparent to us, but at high magnifications it acts like a river of water in continuous movement, continuously distorting the images. The boiling that we observed is mainly due to the movements of air in the firsts 6 miles from the surface.

Atmospheric turbulence, which is parametrized with the word 'seeing', disturbs all images and varies from place to place and from evening to evening. It can't be eliminated but we can ensure that it does as little damage as possible.

Here are a few useful suggestions:

- Never observe from behind a window (it seems obvious to me, but it's best to clarify!);
- Observe the objects only when they are high on the horizon. When they are close to setting or are have barely risen, their light crosses through much more air and, as a consequence, is much more apparently disturbed;
- Avoid observing objects that brush against the roofs of houses because, especially in winter, the houses give off heat as though the roofs were radiators and this ruins the images;
- Cold, windy evenings are usually rougher, while calm nights with a bit of mist are the best because they are stable.

Atmospheric turbulence isn't very obvious under 100 magnifications and considering that I personally use 100 magnifications with telescopes smaller than 250mm (10 inches) diameters only for observing the Moon and the planets, we can devote ourselves in any case, even in evenings when the air obviously boils, to all of the deep sky objects.

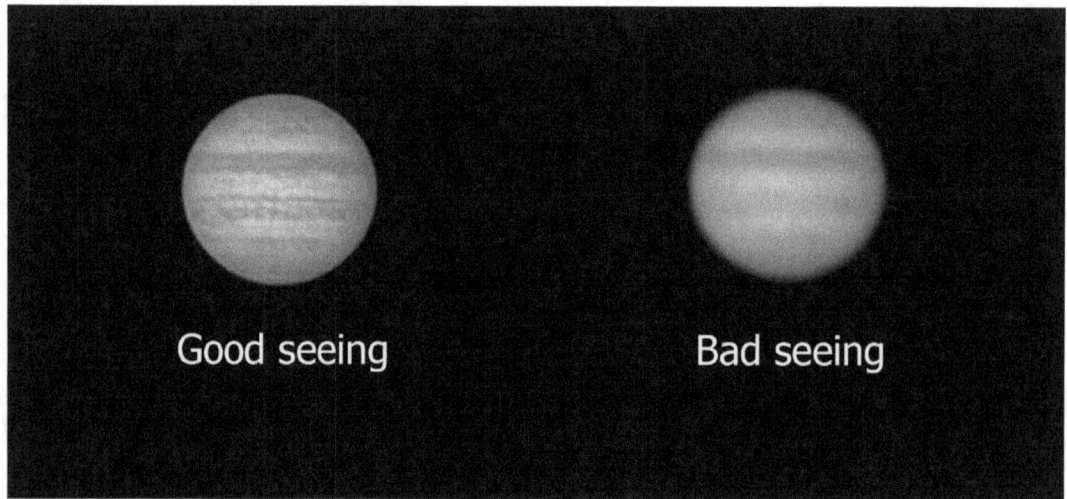

If we observe at high magnification and the image seems to boil, we have just met atmospheric turbulence. It's best to make a few small adjustments to push ourselves, at best, toward the limits of our telescope.

We've probably tried, curiously and ingenuously, to point a few stars with the telescope. We'll have even tried to magnify the image as much as possible, hoping to be able to see the shape, but with uncertain results. This is a great lesson that we have learned through our own efforts: **the stars are so far away and small (compared to other objects) that they can't be magnified** with any instrument, not even with large professional telescopes. So, the objects to be observed are the ones we have already begun to see with binoculars: star clusters, nebulae, galaxies and, of course, the planets and all celestial bodies with a diameter appearing sufficiently large enough in the sky that they can be easily seen through every telescope.

The astronomical coordinates are a more sophisticated way of finding celestial objects, even though we, at least at the beginning, won't need them. However, it's useful to mention them, also because we have already spoken about them when we described that strange thing called the equatorial mount. The ones used the most by amateur and professional astronomers are called equatorial coordinates. We're not interested in understanding in detail how they are measured; a simple comparison is enough for understanding what they are. If we remember correctly, the movements of the sky are the reflection of the Earth's movements. Extending this property a little further ahead, we reach the definition of astronomical coordinates. In fact, if we have latitude and longitude at our disposal for finding a place on the surface, why not do the same thing in the sky?

So, the equatorial coordinates are nothing more than the concepts of latitude and longitude applied to the celestial sphere. To distinguish them from Earth's coordinates (otherwise, imagine the confusion!) astronomers have called them Right Ascension and Declination, abbreviated with RA and Dec. The Right Ascension indicates the longitude of the stars in regards to a "meridian" of reference. Our reference on the Earth is the meridian of Greenwich; in the sky, it's a point in the Pisces constellation called gamma point.

The declination is the analogue of the Earth's longitude and measures the star's elevation in respect to the celestial equator, the projection in the sky of the Earth's equator. In this manner, it's as though the entire celestial sphere is wrapped in a tight grid of coordinates that allow us to find even those objects that don't have their own name or else move rapidly among the stars, like asteroids and comets, without a possibility of error.

The axes of the equatorial mount are called right ascension and declination exactly because they follow this system of coordinates.

Observing the Solar System

The planets have always been my passion. I don't know exactly why, but thinking about observing worlds similar to the Earth, discovering small details inside their discs and tracking them down without problems even with the naked eye gives me a unique thrill that I can barely experience with other celestial objects. It's true, nebulae and galaxies are much bigger, maybe even more beautiful and certainly impressive, but the planets are something more familiar, closer, more easily reached with our imagination. I've made so many trips in my mind by simply thinking about observing a planet that evening with my telescope. And even if I was usually unable to because I didn't know where they were, I remained fascinated by what could have been. Maybe we're not used to doing it anymore, but the imagination – that thing which, at this point, TV, mobile phones and video games continuously try to eliminate – is something that no technological contraptions can worthily substitute.

After having admired the Moon with binoculars and with my jaw hanging open for who knows how many days, my objective – after convincing my parents to give me my first telescope – was to find the planets.

I remember two pages in the instrument's small manual where they briefly described the clearly visible planets and how to observe them. The first was Venus. They wrote, "Visible after the sunset or before dawn". I looked for it for a long time and yet, I could never see it. Then, there was Mars, which, however, could be clearly seen every two years and I didn't know if it was visible or not in that period. Jupiter and Saturn were certainly the most interesting. The first because they said that it was big and clear with the telescope, and even surrounded by four moons. The second because it showed its rings, which that little book said were fantastic with any telescope. Everything was very beautiful; too bad I didn't know where to find them (and, let's be truthful, I wasn't all that interested in observing with the naked eye; wrong! But, I've already mentioned this story and it's better to just move right along.).

I remember that one day in February 1994, in the morning before leaving for school, I made my mom help me set up the telescope in front of the open window in the living room (dreadful!) because I wanted to observe that very bright "star" that could be seen in the sliver of sky between the roof and the tree facing it. Who knows; maybe it's a planet, I hoped to myself.

By then, I had made so many attempts that I truthfully didn't expect to see much, but the excitement of that small hope of finding something different in front of me, compared to all the other anonymous stars I had observed up to then, was palpable. The morning sky, those stars that were so different from those of the evening, fascinated me and maybe hid something beautiful from me.

I pointed the star with the finder, which I had learned to align, at least; I inserted one of the two eyepieces and observed it. I tried to focus on it but that dot continued to stay inexplicably rather squat. So then I changed eyepiece, still without knowing how to calculate the magnification, and I discovered something unexpected. I couldn't bring that dot into focus because it wasn't a dot. It was spherical, contrary to all the other stars, and it had three bright stars around it. I looked incredulously at my mom and told her to take a look because I still wasn't sure what it was and she told me that it might be a planet. Yes, I had done it. That bright star was in reality a planet! I began to distinctly see in that disk two darker spots that crossed it from one part to the other. It was Jupiter, it couldn't be anything else. There, in front of my eyes that glistened with emotion, I had finally observed another world; I had seen in direct, with my own eyes, that which I had dreamed of for many nights over the small manual of the telescope. I remained breathless for who knows how long. If my eyes could have taken photographs like a camera, I would have taken millions of them. But even without film or modern digital cameras, I took them so well with my mental camera that now, 24 years later, I still remember that vision as though I were still observing it with my telescope. Fantastic.

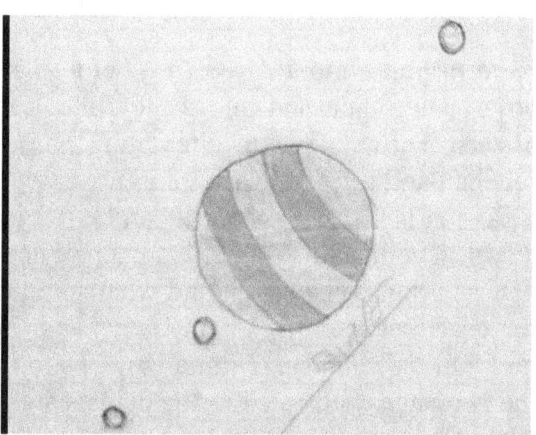

The happiness at having seen Venus' thin crescent for the first time pushed me to draw that wonderful view that I'd had. In the upper part of the drawing, attempts made with a pen are also shown that are possibly even better than the main figure! But who cares about aesthetics. Every time I look at these ugly drawings I relive that unique atmosphere as an explorer of the entire Universe.

My first drawing of Jupiter after rediscovering it years after the first and only observation. I don't know precisely what I saw because now I realize that the position of the satellites and bands (which I also drew crooked) isn't compatible. For anyone who would like to look it up, here's the date: Saturday, November 22, 1997, around 6:30 pm.

Jupiter is the easiest planet to observe and is also the one that gives the best details. However, three more long years went by after that fantastic experience before I was able to observe it again in November 1997, in the sunset-lit sky, near a star that I could finally see clearly: Venus.

How could I not find it in the previous years? It's even brighter than Jupiter! Evidently, because of some strange stroke of bad luck, when I looked for it after the sunset, it was visible at dawn (or wasn't visible at all) and vice versa. Instead, that evening the clear sky and a sudden desire for astronomy after many months of abstinence let me see it very clearly.

I pointed the telescope knowing what to expect, but the surprise is relentless, even when you're prepared. That luminous dot showed up as a thin and spectacular crescent. By then, I knew which eyepiece to use to get the maximum enlargement so I immediately inserted the one with a focal equal to 8mm and enjoyed the view. It was cold outside and I had positioned the telescope in the open window again, this time in my bedroom and without, naturally, acclimatizing it; and yet, Venus bewitched me to the point that for the very first time I decided to bring a pencil and notebook and draw it.

However, Jupiter was there nearby and I couldn't remain on the planet for more than 5 minutes.

My first alien world was more beautiful this time than it was the first. The bands on the globe were clearly visible, the moons were perfectly aligned. I did a drawing of him, too; it's still here with me and with a drop of pride, I'd like to show it to the whole world.

Several more months went by before I used my telescope again. You know, little boys have a thousand things they like, a thousand things to do and it's normal that passions move along in alternating phases, since no one forces anyone to follow schemes and rules, at least not in my case. In fact, there's one thing I'm sure about: if a passion is a true passion, it can be left in suspension even for years, but sooner or later it'll come out stronger than before and we won't be able to do without it.

During the hot summer (and what summer isn't hot?) of 1998, the calm nights and the chats between my grandparents and their neighbors under the starry sky of that house in the country, reignited my desire for astronomy. I picked up that manual and the two pages dedicated to the planets and I realized that I had to observe at all costs the most spectacular of them all: Saturn. I didn't have updated maps yet and I didn't know where to find them. There was no internet, I didn't know the astronomy magazines yet, and so there was nothing left to do but search randomly. If Jupiter and Venus were very bright and I had found them that way, maybe I'd have had the same success with Saturn.

For several evenings, my mom and I tried to look for the planet with the rings. I didn't know what to expect. The small manual said that they (the rings) were very evident with any type of instrument and spectacular at a magnification higher than 50. I could reach 70; maybe it was even too much!

I'm a little embarrassed now to admit that, for a couple of times during my random wanderings among the stars above the eastern horizon, I mistook the reflections of the power lines in front of the house for the magical rings of Saturn.

"There it is, I saw it, I saw the rings!... But now I can't find them anymore; I saw them for a minute. But, how big are they?" I asked my mom, who didn't know how to answer but decided to stay there and help me as much as she could. I looked and looked but there was no trace of Saturn. I decided to look at the early morning sky, from our house in the city to where we had moved a year and a half earlier and in which we spent every day of the week except for the weekends.

The morning of August 2, 1998, without waking up anyone, I woke up at 4:35 to the ringing of the alarm clock and went to the balcony where I had left my trusty telescope leaning on its wobbling tripod.

How different the sky is and was at daybreak. The streets are deserted, even in the city; the only thing to speak is the stupendous silence of the Universe, which seems to have reconquered even this strip of land tortured by man. The air is fresh and calm, even in summer, and the stars are all different from the ones we said good night to just a few hours earlier. That magical night was no different… After a moment of confusion and contemplation, I decided to get a move on. This time, things were easier. Yes, because in the city, the artificial lights only let us see the brightest stars. Considering the fact that from the balcony only the south-southeast horizon was free, I didn't have the temptation to look around too much and get lost again.

In that sky, way above the horizon, I saw two bright stars right away. The first, more to the right, was surely Jupiter because it was far brighter than any star and perfectly still. I pointed it with the telescope and found it in beautiful shape, so much so that I even lost ten minutes admiring it.

However, my objective was Saturn. I had only one possibility: that other star, not nearly as bright as Jupiter, but more so than all the others that I saw in that slice of sky. I pointed it without expecting anything, just a question of healthy curiosity.

As soon as the telescope framed that dot of unmoving brightness, my life changed. Suddenly, my hands were covered with sweat, my heart began to pound like crazy, my mouth was smiling, my eyes popped open as though to gather as much light as possible, trying to understand whether or not this was a dream.

That little star that I hadn't trusted was Saturn itself. The planet I had dreamed so much about and looked for was finally right there in front of me! I couldn't believe it and I couldn't believe how much perfection was right there in front of my eyes. Those small rings, tightly wound around a pale disk seemed to be a vision too beautiful and too perfect to be true. More than once I moved my eyes away from the eyepiece and looked toward that dot disguised as a star. "How can you, little dot like many others, be so beautiful? Am I dreaming or is it real?" I continued to repeat to myself. In the end, my joy was so uncontainable that I went in front of the window in my mom's bedroom and woke her up without disturbing my father. I dragged her, with my boundless joy, out in her pajamas and forced her to admire that miniscule planet with the rings. And even she, who knew nothing about astronomy (like me) and whose ups and downs of life had kept her away from the luxury of raising her eyes to the sky and asking questions, was excited and speechless in front of the vision of the real Universe.

Because the sky is so disarmingly beautiful that when it reveals itself to us, it's impossible to turn away as though nothing had happened. This is the reality, this is the environment in which we live, this is the space from which we come and this will be the future after our time here on the Earth is over. Everything else is frivolous and superficial. And I, that night, could finally shout out my victory to the entire world: OH WOW, I SAW SATURN!

And so my occasional desire for astronomy was transformed into a passion that I would have cultivated from that moment on. The spring that convinced me about the absolute priority of the Universe over the rest of my life was released that early morning in which I couldn't fall back to sleep until the following evening.

I've never lost those planets from view, from that day on. People ask me continuously how I can recognize them and I tell them what I have also written in this book: that I know the constellations and I know when there bright dots that shouldn't be there. Instead, the truth is that the emotions of that indescribable morning were so strong that I was able to establish a tie so strong with the Universe and those two planets that I would be allowed to recognize them anywhere in the sky, even years later. I don't know how to explain how it's possible, but maybe there's no need because it is simply one of the many meanings of love.

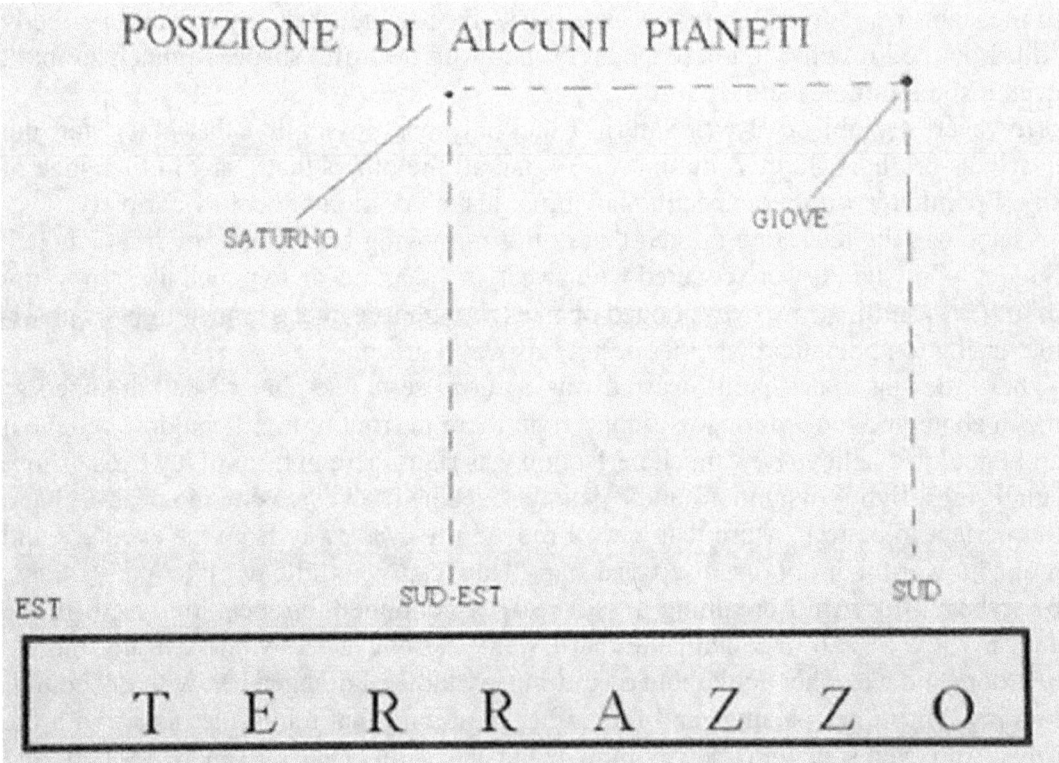

What a view… August 2, 1998, 4:45 am, I observed Saturn for the first time with the telescope. My uncontainable joy for the wonder pushed me to live astronomy as the greatest passion of my life; in fact, it is my reason for living. I woke up my mom and said, "Oh wow! I saw Saturn! Come see!" And so I wouldn't forget the positions of the planets, I drew them that morning on the computer and printed them out, but I never needed this map because I never lost them from my skies or my heart (sorry, the sketch is in Italian!).

Now, many years after that night, every time I observe Saturn it's like seeing a perfect love again which nothing and no one can ever spoil. And when I proudly show it to curious people during the public evenings, and unbelievers ask me if I have put a photograph in front of the telescope because it's so perfect, I answer smiling with a truth that is as simple as it is amazing: the Universe is more beautiful than any photograph ever taken or conceived by a human being.

What can we observe on the planets and how?

I'll admit it: I'm getting a little nostalgic while writing this paragraph because it will be similar to those two pages in my telescope's manual that kept me company for years, letting me travel faraway in my imagination. So therefore, I'm touched to think that someone reading these pages might possibly feel what I have experienced during my voyage from child to man. Personal feelings apart, let's take a good look at the planets to be observed and what there is to see.

Mercury is the smallest and closest planet to the Sun. It has a greyish surface; it is almost completely without an atmosphere and is deeply furrowed far and wide by large craters caused by the impact of large meteorites. From its surface, the Sun appears three times bigger compared to how we see it on the Earth and its temperature reaches 842°F. However, at night, without atmospheric protection the temperature precipitates to -148°F. No one could live on Mercury... too bad.

The apparent diameter is always rather small, on an average around 10". Because it's an internal planet, it shows the phenomenon of phases. We'll see it as a scythe of varying widths, depending on its position in regards to us and the Sun, somewhat similar to a mini-Moon. This is all we'll be able to observe with a telescope, even with the most powerful instruments. In fact, the difficulties in its observation make almost any efforts to see it useless, even at its best. In fact, as it is always low on the horizon it suffers from considerable atmospheric turbulence.

The phase is visible with any instrument starting with an 80mm refractor and around 100 magnifications. It's useless to go beyond 200X because the image will always boil.

Venus doesn't show many details but it's very beautiful to observe. Drawing by Giorgio Bonacorsi, observing with a 130mm (5 inches) Dobson telescope at 220X.

Venus is much brighter and definitely moves further away from the Sun. It's an internal planet and therefore it also shows phases which are visible even with binoculars at barely 10 magnifications. Almost as big as the Earth and much closer (in fact, it's the closest planet, up to 24,854,847.689 miles away!), it shows a minimum diameter of 10" when it is on the other side of the Sun. During its longest elongations it can be observed fairly high in the sky after the sunset or before dawn with an average diameter of 30". Little by little, as its rotation around the Sun continues and it contemporaneously draws closer to the Earth, its angular dimensions grow and the phase grows thinner, becoming very suggestive.

It's a very easy target with any telescope, even with a small toy with a diameter of 60mm (2.4 inches). It's very bright and therefore supports the higher enlargements very well.

Unfortunately, because of its thick, impenetrable atmosphere, always covered with clouds, it's impossible to observe its surface or any details. Only expert observers, with telescopes of at least 150mm (6 inches), are able to extract some faint nuance of the clouds when the phase is at about 50%. So, it's an easily observed target, but gives little satisfaction. Its clouds not only obscure the view of the surface, but present the planet with an intense greenhouse effect that makes the average temperature rise higher than 842°F, even at the poles and at night. Therefore, Venus has earned the title of hottest and least hospitable planet in the Solar System, a true furnace able to melt some metals like lead and solder, an enemy of any form of life.

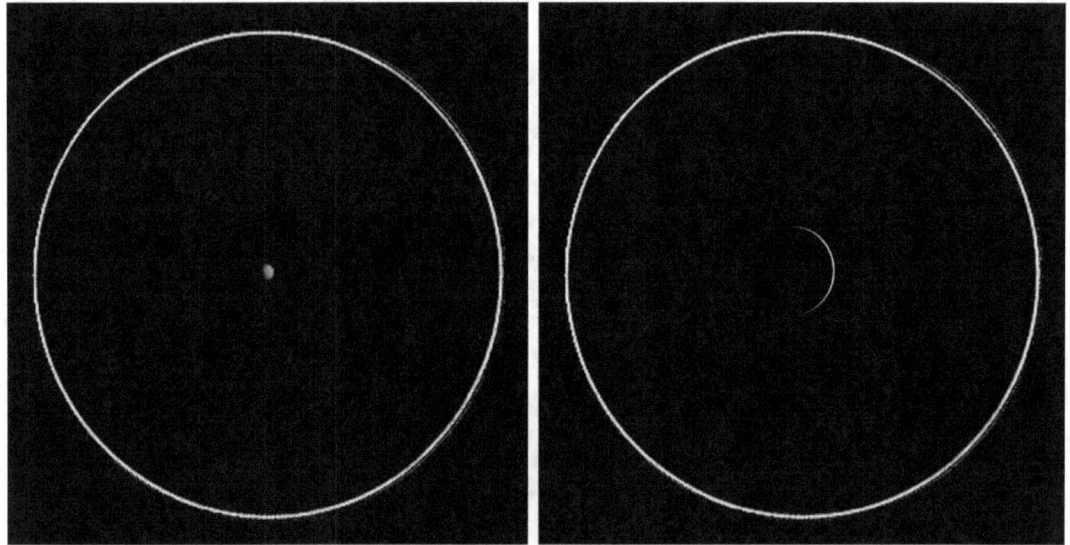

Mercury, at left, and Venus, to the right, as they appear on a telescope of 80-130mm (3-5 inches) at around 250X. Other than the phase, nothing else can be seen.

Mars is the most similar to the Earth and also the most interesting, but it's also among the most difficult to observe profitably. It's not very large and has a slightly lengthened orbit, so it's best observed only when it's closest to the Earth, at the turn of the period called opposition (a term we've already heard a few pages back). The window in which the red planet begins to show details, opens around three months before up to three months after the opposition. Unfortunately, these are rare events that take place every 26 months and aren't always favorable.

In an average opposition, the angular diameter of Mars is 18-20" and this means that to see it as large as the full Moon with the naked eye, we have to magnify it around a hundred times.

It won't be enough, as we already know, but at 150-200 magnifications, if we use telescopes with at least 90-100mm (3.5-4 inches), we can have a nice view. The reddish color will be evident, while we'll notice, at one of the poles, a white polar icecap made of ice composed of water and carbon dioxide. Its tenuous atmosphere allows the colors of its surface to be seen wonderfully. The most beautiful and con-

Mars is the planet most similar to ours. Clouds, fog, polar icecaps and an immense desert that brings the Sahara to mind, even though it's a bit cooler.

trasting detail is Syrtis Major, a long strip of dark land that stands out, when visible, at the center of the small disc. Another detail that is also within reach of small telescopes is Sinus sabeus, another strip, running horizontal this time, not too distant from Syrtis Major. At certain times, telescopes with a diameter of at least 150mm (6 inches) may reveal small off-white spots: they are the clouds of the red planet, very much like the earthly cirrus clouds. The view becomes really beautiful with a 200mm (8 inches) telescope used at 300 magnifications in an evening when air turbulence is very low.

Soft chiaroscuros stand out on the globe and maybe we'll also have the impression of seeing the famous canals of which astronomers of the 1800s believed to have caught sight. Actually, they were suggestive optical illusions that remind us that, at some time in the past, water ran on that faraway world and that maybe there was also life. Now, instead, we'll have to make do and im-

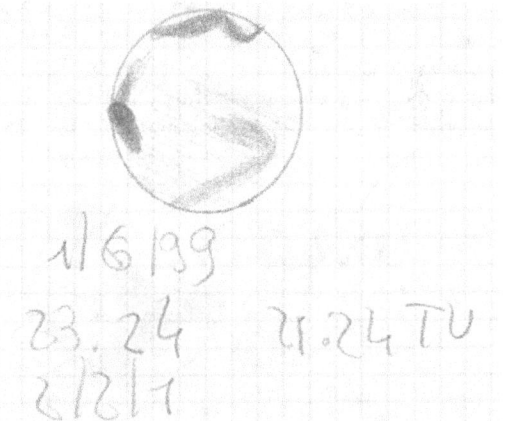

Even though I am very bad at drawing, during the opposition of 1999 I entertained myself for several weeks by trying to extract all the weak details of Mars with my 90mm (3.5 inches) f10 refractor. In the observation of the planets it's necessary to annotate the time in Universal Time (UT, *Greenwich time*).

agine, while we look at it, that some of the automatic spaceships that left from the

Earth months and years earlier are around and on its surface. Someday, maybe, man will be able to stand on it and discover many mysteries that it still guards jealously.

In fact, a very, very long time ago Mars had to have been covered by water and rivers and experience temperatures very similar to those on the Earth. Then, all of this disappeared; the paradise was transformed into an arid, cold desert where not only is there no water anymore, but there are many zones where it just can't exist. If we were to open a bottle, the water would explode like when it is poured on a fiery-hot frying pan.

No living species that we know of could live at this time on the surface of Mars. There's no oxygen, it lacks water, and the atmosphere is too thin and doesn't block the damaging solar radiation. The average temperature is -90.4°F! It's true that at the equator and in the sunniest days reaching +68°F is possible, but then it precipitates down to -90°F at night, and we truly wouldn't want to be around there without an appropriate protection!

A decidedly better drawing of Mars than mine on the preceding page, drawn by Giorgio Bonacorsi during the 2012 opposition while observing with the ocular of a small Dobson of barely 130mm (5 inches) diameter at circa 220 magnifications.

The small Sojourner rover was the first manufactured object to move on the surface of another planet, in 1997. We see it here at work, recorded by the mother capsule which arrived on Mars with the Mars Pathfinder mission.

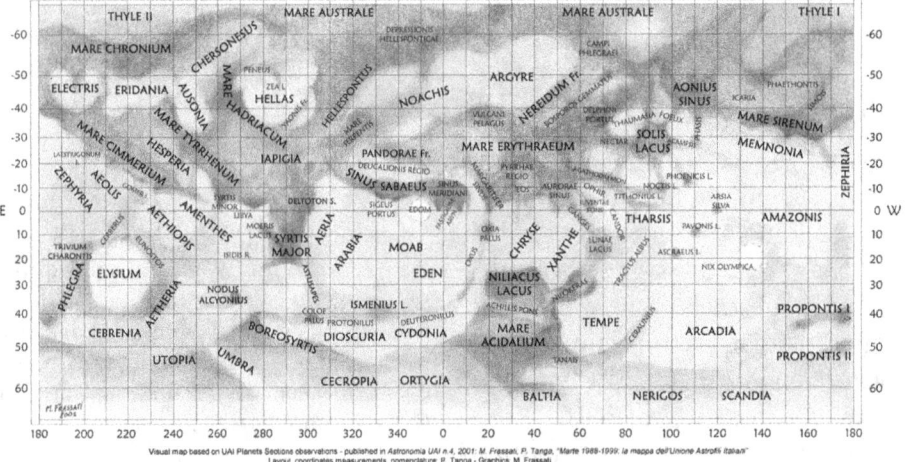

Map of Mars compiled by the Union of Italian Astrophiles (UAI) thanks to the observations of the amateur astronomers. The map shows all the details that a 150-200mm (6-8 inches) telescope can actually give to an experienced observer, in good atmospheric conditions.

Jupiter is the giant of the Solar System and can be observed for about 10 months of the year, every year. The minimum diameter at the maximum distance doesn't go below 30" and in the most favorable oppositions exceeds a little over 50". This means a big celebration for our telescopes and an explosion of easy details which increase with practice. Since it is a gaseous planet, it has no solid surface but the most violent, fast and colorful phenomena of the Solar System are unleashed in its atmosphere. Its two equatorial bands can already be seen with a 60mm (2.4 inches) telescope With a well-trained eye is it possible to see even the great red spot, a cyclone twice the size of the Earth, which has raged for 400 years with winds that reach over 300 mph! Its pale brown color becomes increasingly more evident and tends toward pinkish as the diameter of the instrument increases.

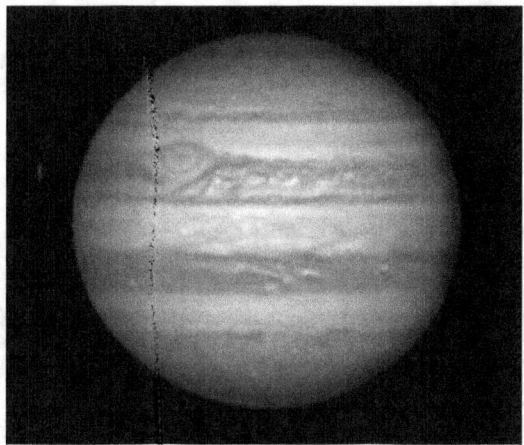

Jupiter and the great red spot. An easy and splendid target to observe with any instrument.

With the same modest instrument, it is possible to witness the dance of the four main satellites discovered by Galileo between 1609 and 1610: Io, Europa,

An eclipse on Jupiter produced by one of its four main satellites. Drawing by Giorgio Bonacorsi done while observing with a 250mm (10 inches) Dobson telescope.

Ganymede (the largest of the Solar System) and Calisto. A telescope with a 90mm (3.5 inches) diameter at 150X gives an already beautiful view, with the planet already three times bigger than the full Moon and appearing oval because of its rapid rotation around its axis (less than 10 hours!). In the disc, other than the two main bands, many other elusive details can be observed, especially between the two bands, in the equatorial zone. We're admiring the famous festoons, atmospheric currents which are seen, depending on the period, as more or less evident and colorful.

One never gets bored with Jupiter, especially with instruments starting at 150mm (6 inches) and up, using a magnification of at least 200. Small cyclones, some of which are colorful and others, clearer; Galileo's moons which, looking closely, appear as tiny diskettes and no longer as dots. It's an explosion of details that deserves an extensive observation and a beautiful drawing. If we're lucky enough to look calmly through a 200mm (8 inches) instrument, we'll see even more details, with the colors even more evident and very much like an extraordinary painting.

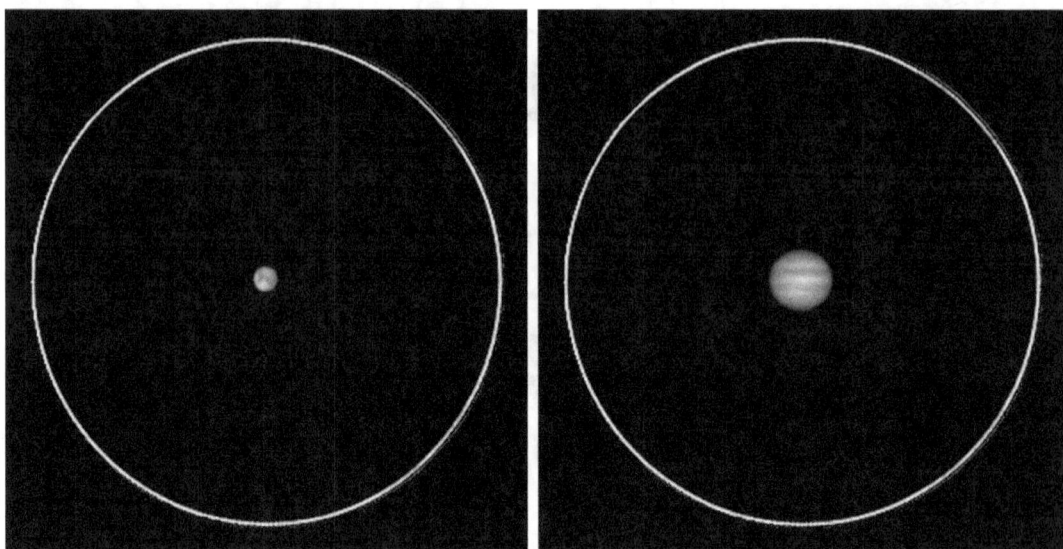

Mars, at left, and Jupiter, at right, observed the first time with a small 100mm (3.5 inches) refractor at about 200X. Although our brain will make us seem them small and with very few details, we'll see them better than this as time passes.

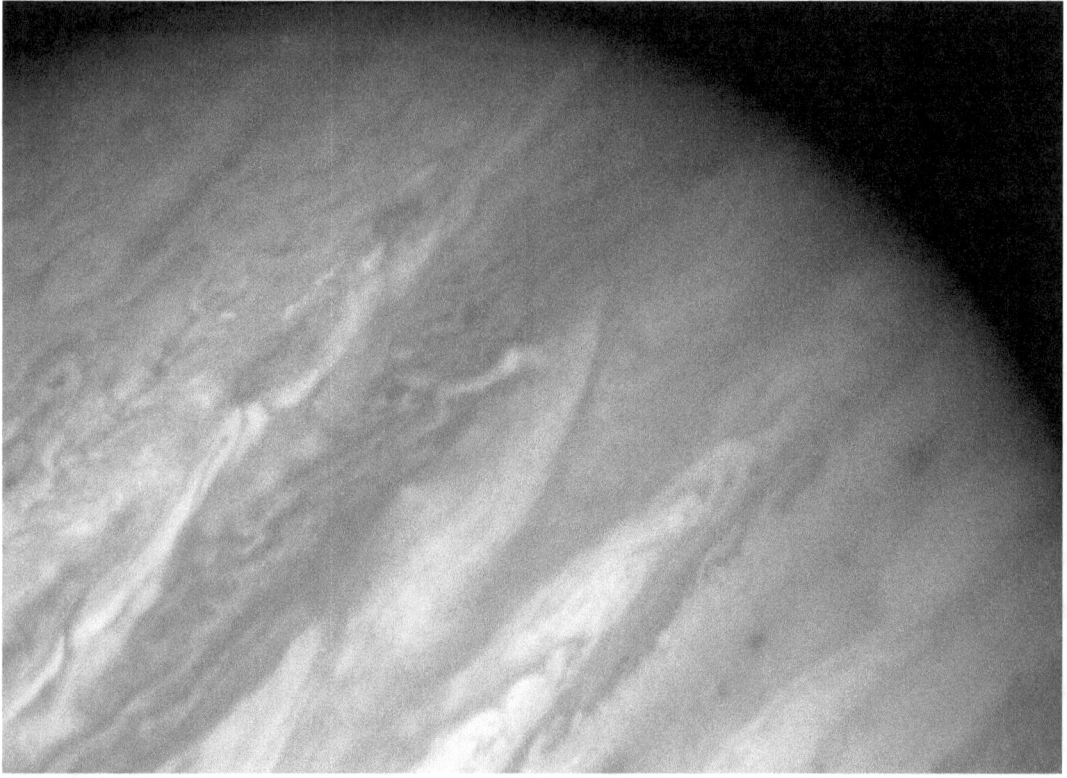

Impressive image of Jupiter from the Hubble Space telescope.

Saturn is the lord of the rings, clearly visible with any telescope starting with 15-20 magnifications.

The beauty of this planet is indescribable, so perfect that at times it's difficult to believe it really exists. Much calmer than Jupiter, because of its greater distance (almost 746 million miles from the Earth!), it's easy to observe with any telescope. The globe's diameter is about 19" while that of the rings reaches 45". Because of the enormous distance, it can always be observed very well in any period of the year, as long as it visible and high above the horizon. Instruments higher than 70-80mm used with at least 100-150 magnifications permits us to notice a dark thin line in the rings. This is the famous Cassini division, a 2500-mile wide gap inside of which the rings are much more rarefied.

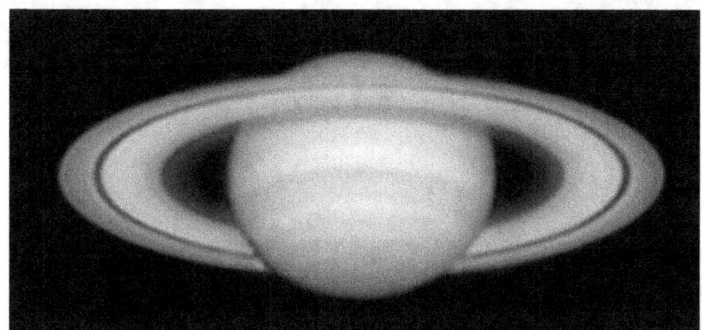

Saturn is a unique planet that thills you every time you decide to admire it.

Constituted by crushed stone and powders and, to a large extent, composed of water ice, they are only a few hundred yards thick and, therefore, don't appear homogeneous.

A telescope of at least 100mm at a 200X will allow us to notice some bands also in the globe's atmosphere, in addition to several small, very interesting details, like the planet's shadow on the rings and at least 4-5 satellites. The brightest, Titan, can be observed with any instrument, while the others, fainter, require a bit more attention because they are always close to the planet's bright profile.

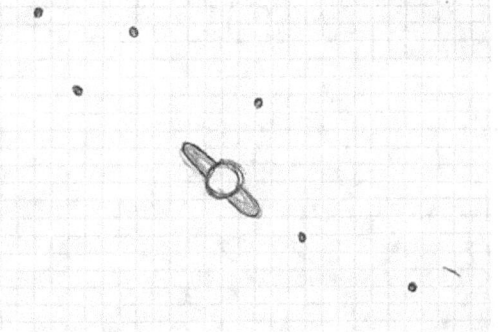

The first observation of Saturn certainly can't be forgotten; the rings are incredible to observe, even if they are hard to draw! Refractor telescope, 80mm (3 inches) f5.

Instruments of 150mm (6 inches), at 300 magnifications, if the evening permits, will give breathtaking images that are more spectacular and suggestive than any image we can look at on the internet. Let's savor this sublime moment by taking a lot of photos with our mind and trying to retain every little nuance of this gigantic work of art that floats freely in space.

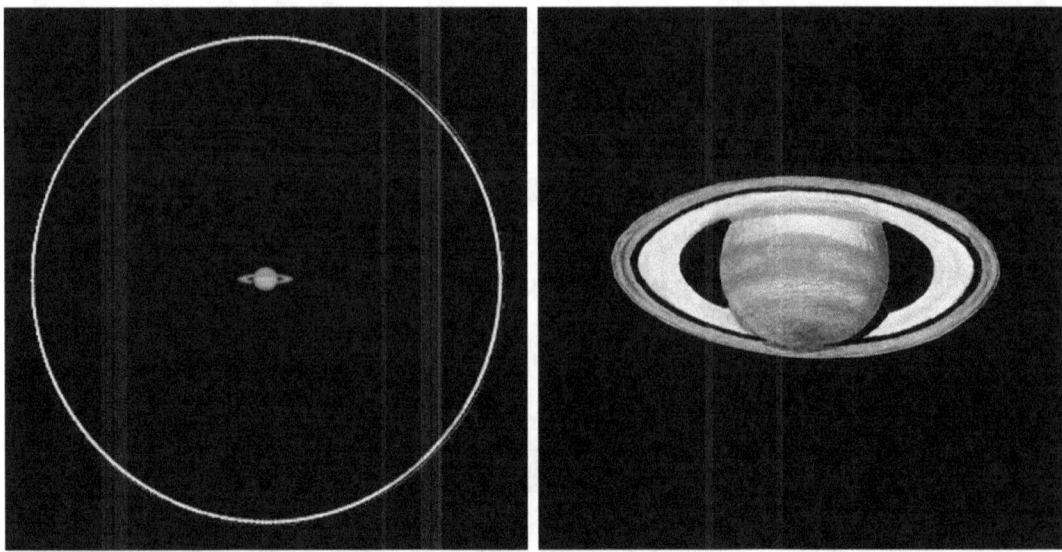

Magnificent Saturn and its rings. At left, how they appear through a modest 80mm (3 inches) refractor; at right, with an instrument of at least 100mm (4 inches), after gaining the necessary experience. It's an exceptionally beautiful view.

Saturn's system isn't beautiful to see just with a telescope, but also to explore with a spaceship like the automatic Cassini probe, which has photographed Titan, the main satellite, and part of Saturn's globe on which the shadow of its rings is projected.

Uranus and Neptune are invisible to the naked eye, so they are hard to find. Their diameters appear very small, averaging 3.5" for Uranus and only 2.5" for Neptune. They are easily seen with every telescope, but the small discs can be seen only with instruments of 80mm (3 inches) up, pushing the magnifications up to 200X. With 100mm (4 inches) refracting telescopes or 150mm (6 inches) reflecting (this being more or less the relationship between the visual performances of a good refractor and a reflector) their colors can be easily noticed. Uranus is pale green, while Neptune is light blue. No telescope is able to discern details, but the idea of being able to see a small disc more than 1.89 billion miles (Uranus) and 2.9 billion miles (Neptune) away sends shivers up and down the spine. Often when I observe them I start to imagine how it would be in those dark, cold places, where the sunshine is dozens of times weaker and its disc is almost a dot in the sky. A dark and silent place from which I could observe all of the other planets like miniscule moons and to lose myself in open space. Fascinating…

The Sun and the Moon

Even though they aren't planets, these two celestial bodies are clearly observed with a telescope, but with a caution about the Sun, which I will write in capital letters even at the risk of seeming rude: **NEVER OBSERVE THE SUN WITH ASTRONOMIC INSTRUMENTS WITHOUT AN APPROPRIATE SOLAR FILTER PLACED IN FRONT OF THE OBJECTIVE.** Pointing the Sun without a filter is extremely dangerous and you risk losing your sight even with a fleeting glance. Please, don't do it.

When one is young, one tends to ignore or challenge danger and I certainly wasn't an exception. I knew that I shouldn't observe the Sun, but one November day thick fog shielded its light and made it so that admiring it with the naked eye wasn't annoying and so I decided to point it with my binoculars. Right then, however, the fog thinned little and the sunlight blinded me for a few minutes. After a big scare, I began to see again, but I understood through experience that I should never have done something like that. The Sun should be observed only with appropriate filters that are put in front of the objective, before the light enters the telescope. Never use homemade solar filters like welding masks or x-rays. Our sight is too precious and once it's gone, it doesn't come back.

Having said this, if we buy a solar filter for our instrument (the Astrosolar by the Baader Planetarium company is excellent and economical), we can observe it safely and comfortably. Before doing it, however, let's remember to cap the telescope's finder or to completely remove it because this can also cause damage to our sight.

One day during a normal day of solar observation, I forgot to cap the finder. I didn't use it to point the Sun, but while I was observing with the eyepiece I began to smell a slight odor of something burning and then terrible pain: my shoulder was burning because it was hit by the sunlight projected by the finder!

I could claim that it was because I was just a little boy without experience, but I don't want to tell a lie: it was last year. I was grown up and an expert, but evidently I wasn't careful enough. All this is to show you that you can't play around with the Sun.

To point it safely, put the filter on the telescope, take off the finder and move the instrument while observing the shadow it casts on the ground. When this reaches the minimum extension the sun will be in the framed field and we can observe it.

If a small solar filter to be screwed onto your eyepiece has been provided with the telescope, throw it away now, because it isn't safe. The heat accumulated after passing through the telescope can make a hole in a plastic cap in two seconds, thus breaking this filter and causing serious damage to your sight.

After this very necessary lecture on safety (safety first, as they say), what kind of cool things can we see on the Sun?

Well, first it's better to understand just what the Sun is. We know that it's a star, and so it shines with its own light, contrary to the planets and the Moon which are limited to reflecting it. But this isn't a great answer, because we can always ask: "What is a star?" And there's no escaping this. A star like the Sun is a gigantic sphere of fiery gas, composed by about 74% hydrogen, 24% helium and the remaining 2% are all the other elements that here on the Earth are considered abundant, among which are oxygen and carbon. The surface of a star appears very bright because it's very hot; in the Sun's case it's 9932° F. But where does all this energy come from? From the internal regions where the temperature can reach up to 27 million (!) degrees and where the extremely violent clashing among the particles which compose the gas give life to what is called nuclear fusion reaction. Basically, the Sun and all the other stars are immense nuclear fusion plants, something that we'd really like to build here on the Earth to replace the dangerous and damaging ones based on fission, the opposite procedure.

In the center of the stars, the hydrogen atoms without an electron collide so violently together that they merge, forming an atomic nucleus of another substance: helium. An incredible amount of energy is released during this process. From the fusion of one gram of hydrogen, we would obtain the same amount of energy as we'd get from burning 11 tons of coal. In practice, for us it would mean eternal energy! The matter present in the center of the Sun is enormous, so much so that it makes it shine with an extremely high brightness that will last for at least 5 billion years.

The Sun doesn't have a solid surface, but the highest layer – which emits the light that we see – is called photosphere and it's the one that we observe with the telescope.

On this surface, circa 187.5 miles thick, solar spots can be observed, darker areas with interesting shapes and full of shades, which cross the disc as the days go by. The larger spots are visible even with the naked eye (using a solar filter!) and are splendid with any instrument. They increase in number with the passing of time, reaching a peak every 11 years: it's the famous solar cycle. The solar spots are real chasms in the photosphere, with temperatures up to 1832°F below the average.

Near the border, it's possible to admire areas that are clearer than normal, called floccules. With a telescope of at least 80mm (3 inches) diameter at 100X, a sort of web can be distinguished in the central regions. We're observing the granulation, sacs of gas with a typical diameter of 625 miles, which rise from the depths, releasing their heat and then sinking once again, all within the arc of around 10 minutes.

Using particular telescopes, which we most likely don't have, called H-alpha (or solar) telescopes, it's possible to admire the more violent aspect of the Sun: the protuberances and the flares, nothing more nor less than explosions in the lower part of the atmosphere that project into the space of the spectacular fountains of incandescent gas, as though coming from the mouth of an enormous volcano. This type of telescope is very expensive and dedicated only to the observation of the Sun, so it is therefore best to do what I do: become friends with some enthusiast that has one and take advantage of it to take an occasional free look!

We've already spoken about the **Moon** and there's not much more to add except to have fun exploring – maybe with a magnification of over 150X – the entire surface, hunting for strange, and often hidden, details. Besides the larger craters, which we've already seen on the telescope, imposing mountains and mountain chains, valleys, cliffs, small hills and the particularly interesting play of lights and shadows along the lunar terminator are all evident.

An entire lifetime will never be long enough to explore all of the Moon's nooks and crannies at a high magnification. A good lunar atlas is ideal for learning the names and properties of the formations. I'm not a fan of the Moon, but craters like Clavius, Copernicus and Plato, or mountain chains like the Alps and the Apennines (yes, we're on the Moon!) are very famous formations that are always spectacular with any telescope. If we want to discover something more before buying an actual atlas, we can begin having fun with what I have begun building on my website, which can be reached with this link:

http://www.danielegasparri.com/Italiano/moon_atlas_index.htm

Regarding craters, I don't know who is aware of it, but these are actual holes left by the very violent impact with thousands of asteroids during the course of the Moon's history. In the 4.6 billion years of life of the Solar System, millions of meteorites have also fallen on the Earth, but here, winds, earthquakes, water and vegetation quickly erase these disquieting signs. And yet the Moon is the most evident witness of the huge impact of history on our planet. Shortly after the Earth's formation, a planet as large as Mars, named Theia rammed into us. Theia vaporized during the impact and caused pieces of the young Earth to shoot off into space, uniting together to form the Moon. What a beautiful story, right? The Moon as the Earth's rib. This, too, is the beauty of astronomy.

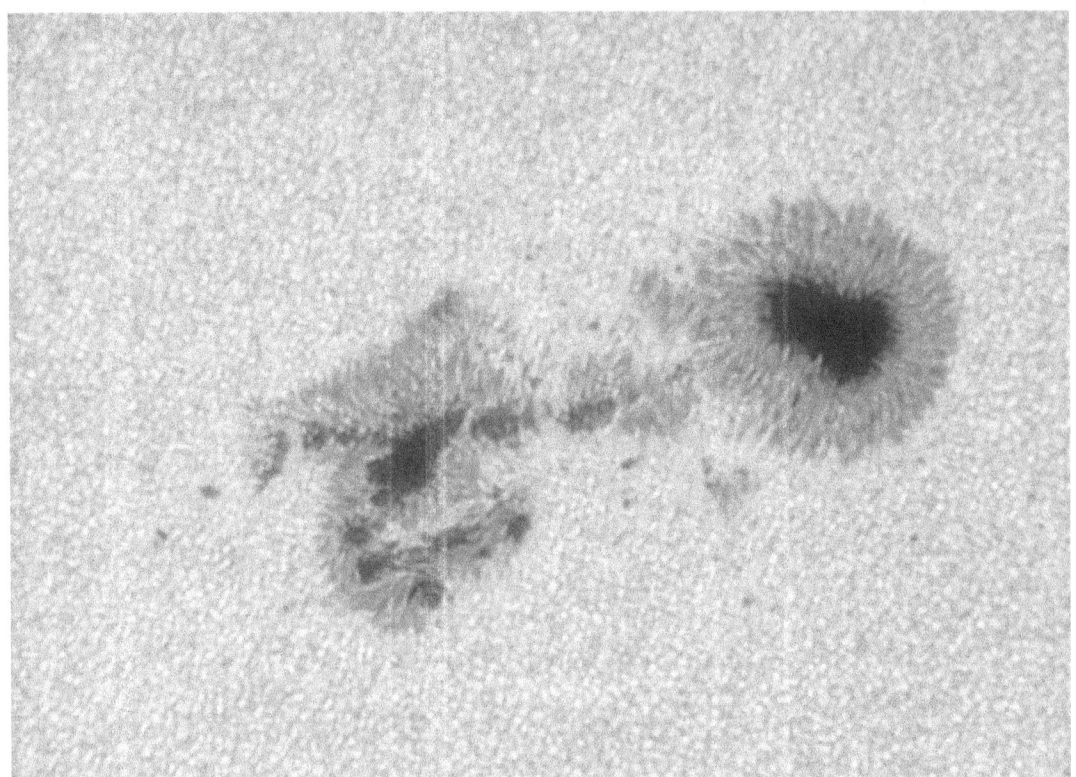

When large solar spots appear on the Sun, observing them with a telescope is truly fantastic. But always use the solar filter!

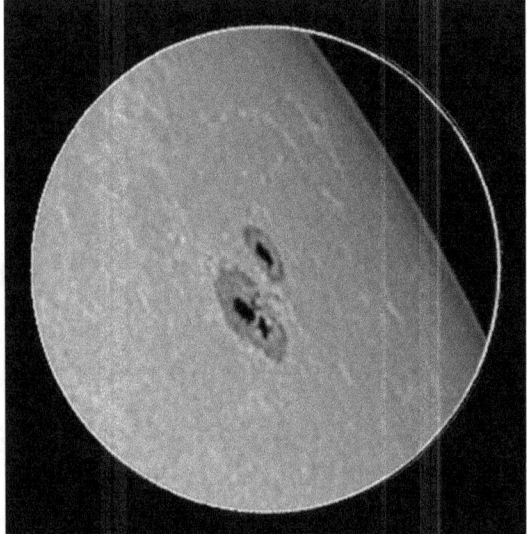

The Sun, through an appropriate solar filter, easily shows solar spots, granulation and floccules, even with instruments having small diameters like 80-90mm (3 inches).

The views offered by the Moon are visible with any instrument, even with modest magnifications.

We can't see either the remains of the spaceship or the flags planted by the astronauts, but we can have fun with our telescope zooming in on the zone where men walked during the '70s. From the first mission (Apollo 11) to the last (Apollo 17) we can fly over the lunar surface at a high magnification and dream of being on board a spaceship that is about to land, just like those fortunate astronauts of many years ago.

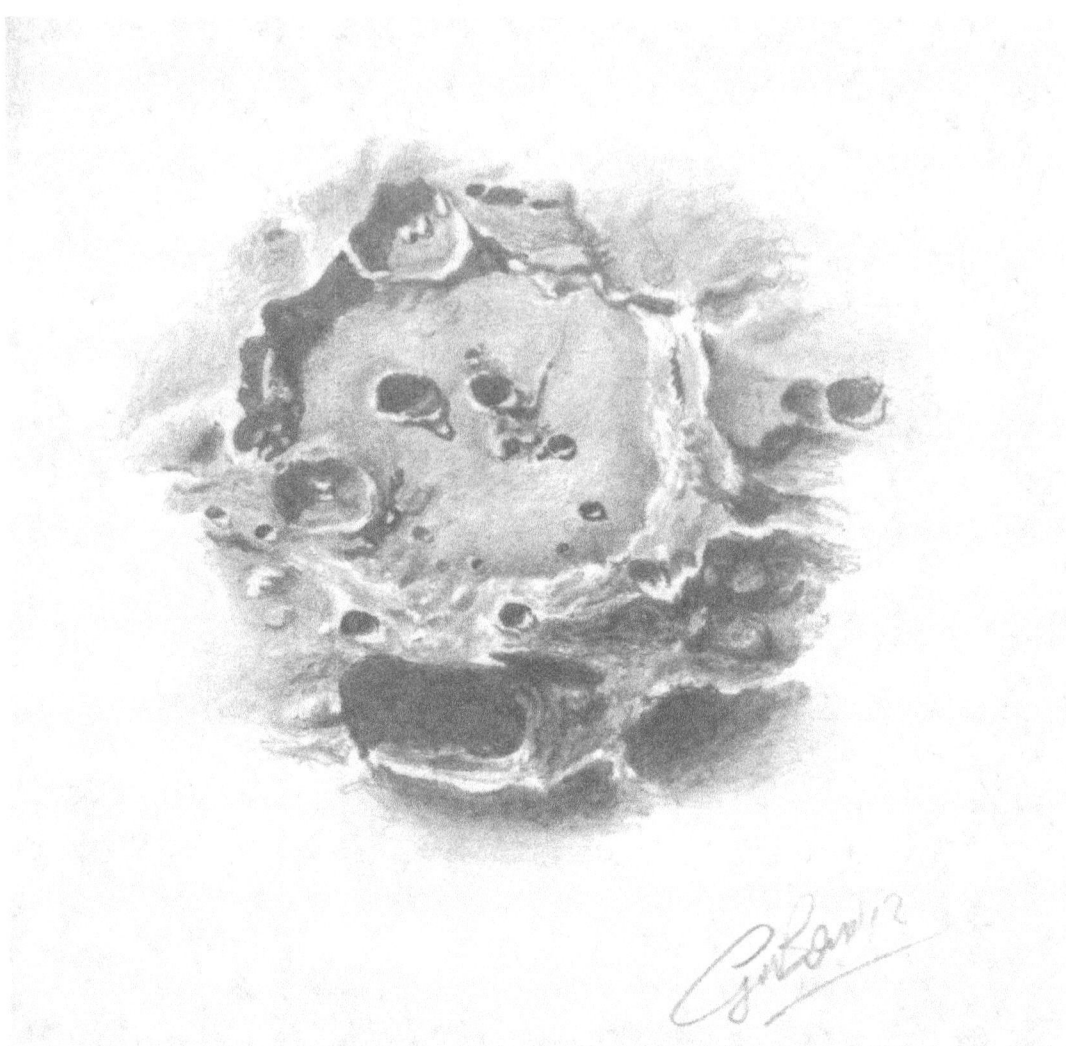

Splendid drawing of the big crater Clavius drawn by the astrophile Giorgio Bonacorsi while observing through the ocular of an 80mm (3 inches) achromatic refractor. The Moon is magnificent seen through a telescope.

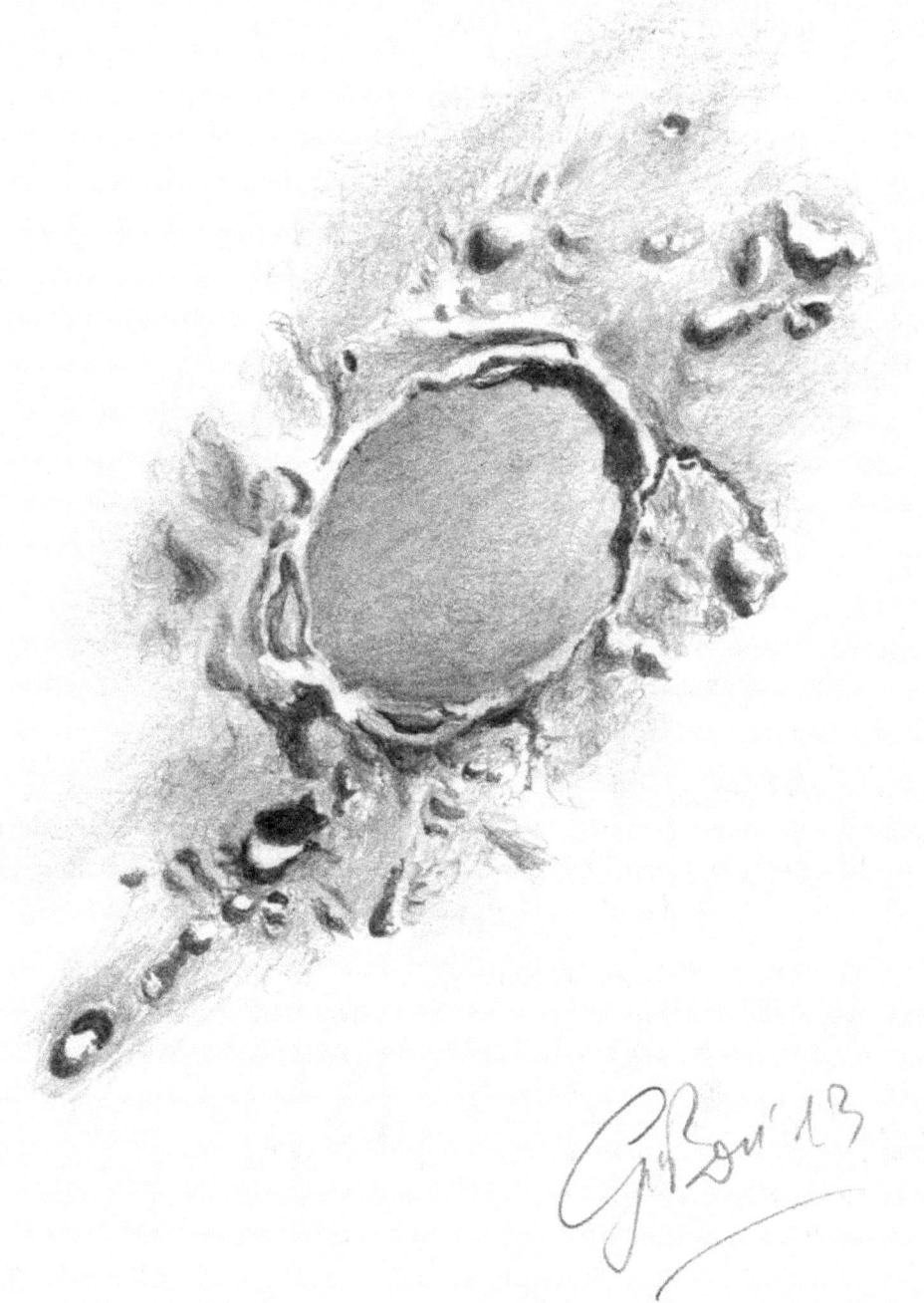

The Walter crater drawn by Giorgio Bonacorsi using the same instrument as in the preceding image.

Asteroids and Comets

They are called minor bodies because they are actually much smaller than the planets; however there is a frightfully high number of them in the Solar System. Asteroids and comets are small bodies, going from a few inches up to a few hundred miles, which have a common origin: they represent the Solar System's discarded matter, pieces of rocks and ice that weren't used at the time of the formation of the planets and which are now found roaming aimlessly around the Solar System.

Asteroids are rocky bodies, often with irregular shapes and concentrated mainly in a belt between Mars and Jupiter, called the main belt of asteroids. Comets have the same properties but are much further away from the Sun and are mainly made of gas and frozen liquids.

When one of these celestial bodies – which orbit almost always beyond Neptune's orbit – is disturbed by some planet or by other bodies, its orbit around the Sun can be changed and projected into the internal zones of the Solar System. When it is more or less at the same level as Jupiter, the Sun's heat is enough to begin melting the surface ice. As the comet approaches nearer to the Sun, the evaporation of the ice becomes increasingly faster. This is when the evaporated gas, together with the particles of surface dust, is dispersed into space for tens of millions of miles, creating the beautiful tail. So, comets are frozen bodies that are slowly melting. Their appearance is extraordinary if they become very bright, which happened, for those of us in the northern hemisphere, with the Hale-Bopp comet in 1997.

I remember perfectly this brightly shining, plumed orb – stupendous, when observed with my little telescope – that kept me company for several months. I even tried to take a few photos, but the developer, thinking that they were shots that had come out badly, didn't develop them. Only years later, thinking back to the negatives, did I realize that the photos had had turned out fine and that the developer was wrong. Unfortunately, I never found the photos again.

Regardless of this small, sad memory, you don't have to wait all this time to admire a comet, because we can observe at least one a year with our telescope. I can't be of use here, because comets are unpredictable and appear when you least expect them, so it's best to stay informed through internet.

Asteroids, instead, are more predictable and some, like Ceres and Vesta, are easy to observe even with binoculars (but just as little specks). From time it could happen that a small asteroid approaches very close to the Earth and becomes visible with our instruments. In this case, there is twice as much excitement, because we won't just see an unmoving speck like a star anymore, but it will be moving quickly. And to think that that dot, so anonymous for those who know nothing about the sky, is actually a cosmic mass that is literally brushing past us at a speed of some tens of thousands of miles per hour. Thrilling...

The Hale-Bopp comet, which appeared in 1997, became one of the most famous and beautiful comets in history.

Advices for better observations of the planets

At this point, we know how to find the bright planets and recognize them in regards to the stars; but if we've been able to see one with our telescope, then we're aware that they may be a bit disappointing: they seem way too small for our eyes, to the point that the details many expert observers say they've seen completely elude us.

When I rediscovered Jupiter in the summer of 1998, I only saw the two main bands, contrary to my trusted manual which spoke of structures that I couldn't find. Now, if someone wanted to challenge me and showed me a planet with any telescope, I would see more details than you would. Am I better? Do I see better? Did I invent them? No, I've practiced! **The details jump out better after a few minutes of careful observation and increase greatly as we gain more experience.** Already after ten observations we'll be able to see much more than at first.

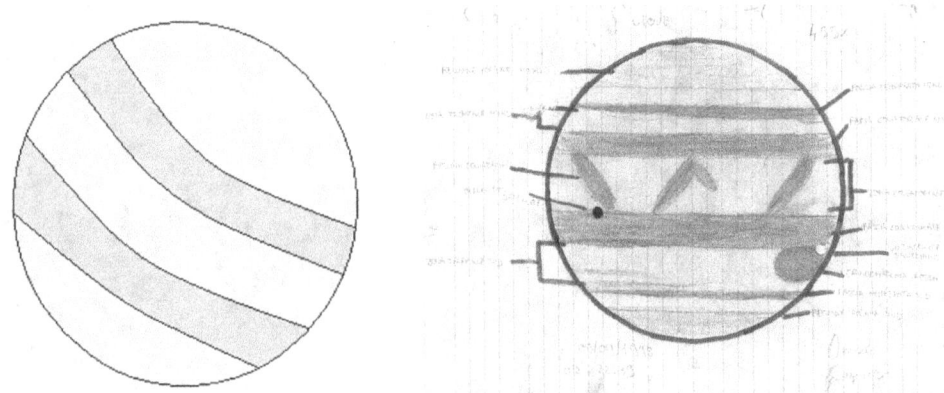

Here's how much experience and practice count in observing the planets. At left, one of my first observations of Jupiter. The drawing, done on the computer, shows no details other than the two equatorial bands, moreover with a rather strange shape that doesn't correspond to reality. At right, the same instrument, a 90mm (3.5 inches) refractor, same magnification, 250X, but a year later and, above all, with more experience. You can't be in a hurry while observing and you need to capture every minimum detail or nuance.

Because I'm a normal person, this is probably the improvement that each of us can expect with a little practice.

However, the fact remains that the planets still seem small from the eyepiece of a telescope. This is a real optical illusion, caused by our brain, which isn't sure how to act when we observe with one eye only one solitary detail immersed in the black sky. With a magnification of barely 50, it appears as big as the full Moon when seen with the naked eye. It won't seem enormous, but it's more than sufficient to see a lot of details. And yet, our eye will find it hard to believe and will show us a planet that seems to be much smaller.

It's not a problem. If it's high on the horizon and there's not much atmospheric turbulence, we can easily see it at a 200X magnification, even with a small 80mm (3 inches) diameter refractor, the smallest instrument for observing a lot of details on the planets. At a 200X magnification, Jupiter appears more than 4 times larger than the full Moon seen with the naked eye. Even if it doesn't seem like it, it's gigantic! If we don't see any details, we just need a little practice. It's unadvisable to magnify excessively. It's better to not buy a second telescope – like a child I knew – just because he could reach a higher magnification with a longer focal, without buying new eyepieces. Too bad the diameter of the telescope was only a centimeter wider; he didn't see any more details. That child – so, what else is new? – was me.

Personal matters aside, it's really cool and fascinating trying to draw what we see on the little planetary discs. It doesn't matter if we're good at drawing; I'm certainly among the worst and yet, this didn't stop me. We're not out to win an art contest but to give the best value to our experiences, maybe even by compiling an actual diary of our observations where, besides our drawings, we write down our impressions of the objects we've observed.

During that magical August of 1998, I dedicated weeks to Jupiter with my new telescope, every evening jotting down the positions of the four satellites and all visible details. Because, if we think that the planets are always the same, we're making a big mistake, as we'll see better soon.

Galileo Galilei's original observations of Jupiter and its main moons. These are among the firsts observations of our history. What about to repeat these sketches, by following the amazing dance of the Galileian moons with our telescopes?

Here is how the planets appear with a 200mm (8 inches) telescope, in favorable atmospheric conditions and having the necessary experience. Above: left: Mars; right: Jupiter. Below: Saturn.

Why observe the planets?

The nice thing about planets is that they can be observed every time that they are seen high on the horizon, even when the Moon is out and even when we're immersed in the city lights. In fact, in these cases we take advantage of only one of the telescope's characteristics: its resolving power (called also resolution). We don't need its capacity to gather light because the planets are already very bright. Therefore, observations of the planets can be made even from the balcony at home, like I do currently.

There are only four nice planets to observe; no, actually three and a half when you consider that Mars can be seen well only once every two years; and yet, there are enthusiasts that observe only these, for years.

I can remember, as though it was yesterday, that every time I went out on the balcony to observe Jupiter my mother curiously asked me what there was to look at. Aren't the planets always the same? Absolutely not. The planets are the most dynamic and surprising objects that we can find in the entire Universe.

Jupiter is the king of transformation. With its atmosphere so rich in colorful cyclones and with a rotation that is completed in less than 10 hours, its appearance changes in just a few dozen minutes. Even the 4 main satellites that orbit around it move quickly, changing their position in the arc of a couple of hours. At times, one of them hides behind the planet and seems to disappear. Other times, it passes in front and almost merges with the giant's big atmospheric cyclones and, not too far off, projects a very dark and suggestive shadow.

Saturn's rings change their inclination every year, appearing very open or almost edge-on with the flowing of the seasons. And if we look at them closer, small white spots called WOS suddenly appear on the globe; these are nothing more than cyclones that disappear into thin air after a little while.

The polar icecap of Mars, just like those here on the Earth, pulls back and advances as the weeks go by. At times terribly violent sandstorms rage and become visible, even with a small telescope, as bright spots near the equatorial zones.

Faced with these frequent events, there are others which are rarer but even more spectacular. And then, suddenly, a planetary storm breaks out on Saturn and is visible for months, just like in 2010. An equatorial band disappears on Jupiter, or else a meteor falls headlong and leaves black marks in the atmosphere for a few months. On Mars, some storms can reach a planetary scale and engulf the entire planet in a gigantic red cloud. And if this isn't enough to glue us to the eyepiece, the extraordinary thing is that 90% of these phenomena aren't discovered by professional astronomers equipped with gigantic instruments, but by us enthusiasts, with small telescopes and an enormous passion!

The deep sky

My relationship with the deep sky has been much more difficult and tempestuous compared to that with the planets. Hoping I don't influence your choices, I'll tell the truth: the objects of the deep sky are gorgeous, but it's exhausting to observe them in a worthy fashion!

My first two instruments had a diameter of 80 and 90 mm (3-3.5 inches) and were good for the planets but pathetic for anything else. Star clusters and galaxies were all the same: dim, indistinct little clouds, even with a dark sky.

Although the planets can be seen with any instrument, the objects in the deep sky require two things: an extremely dark sky and a telescope with a generous diameter.

In order to not run into disappointments, at times even bitter, and to at least distinguish a globular cluster from a nebula, if we are enthusiastic about these objects and have a good sky, **we must have a telescope with a diameter of at least 130-150mm (5-6 inches).** Many expert enthusiasts won't agree with me, but I remember all too well what I underwent with instruments that weren't up to the task. Since at this point Dobsonian telescopes cost less than a cell phone, let's go for the diameter and not worry about the instrument's bulk.

What's the point of observing nebulae, galaxies and star clusters?

When I was able to point the M13 globular cluster from the dark sky at my grandparents' house, called by all of my books the most beautiful in the northern hemisphere, almost as big as the Full Moon and rich with stars, I didn't see any of that. As soon as I identified it I felt extremely disappointed: I only saw an indistinct ball of light without any stars. Ten seconds and then I decided to change objects, but things didn't improve.

A long time later, during a public evening, I could observe it again with a similar instrument (a 90mm, 3.5 inches, diameter refractor) and a much worse sky, and yet, it seemed like something entirely different. Very extensive and bright, I could, with the averted vision technique, see a few dim stars and understand that that "thing" had a granular appearance. Certainly, it wasn't spectacular because of the limited diameter of the telescope, but it was much more evident and detailed than that first distant and disappointing observation.

The moral of this story is always the same, but it's worth repeating because it makes all the difference between a disappointing glimpse and happily rejoicing, regarding the objects of the deep sky. It's all a question of **experience.**

Of course, contrary to the planets, all deep sky objects require **a dark sky**, far from the city and the absence of any lunar crescents; but, training is on a par with these winning conditions.

We've had a taste with binoculars, but in this case it's even more evident. So, we need to be patient and observe the object for a few minutes, searching for details that seem hidden at first, letting our eyes get used to it. Then, little by little, as we train our eye to these strange conditions (which we'll never experience during the day) we'll see more and more, and clearer, until we perceive details that we thought were impossible.

How much experience do we need? It's difficult to give a number and I can't remember when I was finally able to appreciate this type of observation. Probably, if we do it constantly, we'll see a significant improvement already after 4 or 5 evenings.

Another sensitive issue regards magnification.

If we wish to reach the telescope's maximum resolving power with planets, reaching the limit of what we have called maximum useful magnification, the opposite is true for deep sky objects.

When I tried to observe nebulae and globular clusters, I started out with a magnification higher than 100 and could hardly see anything. I thought it was only a problem of experience, but it was more than that.

All the brightest objects of the deep sky are, in fact, much more extensive than the planets, often even more so than the Full Moon. It doesn't make sense to magnify them as though they were tiny discs, and now, perhaps, it even seems obvious, but until one understands this difference, it's difficult to reach this conclusion in a short time.

For nebulae, clusters and galaxies, we have to take advantage of the instrument's other characteristic: the capacity to gather light. So, we need to stay away from high magnifications and eyepieces that seem to be a keyhole. The best magnifications vary from 30 to a maximum of 100 times. Only when we'll be much more expert and want to search for smaller objects (and they are there!) can we gradually experiment fearlessly with higher magnifications. But for now, no, also because the Orion Nebula, the most beautiful of the sky, is more than four times more extensive than the Full Moon, just like the Andromeda galaxy or the Laguna nebula.

If we're curious about having an idea of what to expect from some of the deep sky objects, according to the telescope's diameter and the quality of the sky, you should take a look at a simulator of telescopic views that I prepared a few years ago:

http://www.danielegasparri.com/Italiano/telescope_simulator.htm

Where are deep-sky objects found?

The planets are bright and are seen with the naked eye, even when you're under a street light, so there's no problem pointing them with the telescope (if the finder is aligned!). It's a completely different story for deep-sky objects, many of which are even invisible if we don't use our instrument.

The hardest part of observing deep space regards searching for these dim jewels, a problem that we can and must transform into a challenge to be won at all costs!

The GOTO doesn't represent the solution but is often a blind alley that keeps us from improving our knowledge of the sky and will deprive us of the joys of observing. If we have it, let's decide to not use the first few times. If we don't have it, we'll have saved a lot of money and we'll surely have an instrument with a greater diameter. It seems like I'm being mean, but it's a little like when our teachers at school forbade us from using our calculators when we were learning the multiplication tables. If they hadn't done it that way, we would have been slaves to that technological gadget forever. After we've learned how to do it manually, we can turn, when necessary, to the help of the GOTO, which can, truthfully, help in certain situations and is useful for very dim objects.

When I began to timidly face the deep sky in that long ago August of 1998, I didn't know what to do, but I didn't give up. Finding an object to observe required great concentration and several minutes of studying the sky and trying. Now I can find many objects by heart, without using the finder. Am I a genius? No, simply someone who has practiced a lot. And what can we say about the amateur astronomers who are fond of their wooden Dobsononians that find hundreds of objects much more quickly than the automatic pointing of any instrument? Some call them, with a touch of envy, GOTO humans but they're not aliens, but rather the more obvious witnesses of how our brain is still the best instrument for living life and passions.

In life, in sports and in astronomy, to become good, you have to practice and never become discouraged. Everyone, sooner or later, if they keep going, can reach their goals, an astronomy lesson that also becomes an important lesson in life because, all things considered, the sky is the greatest school of life that we could ever receive. One more reason to love astronomy!

During the course of the centuries of observations, many astronomers have searched the sky and catalogued all the diverse objects, starting from the single stars, far before understanding what they really were. The first to undertake this great work of discovery was a French lord named Charles Messier. With his small telescope, similar to a modern 80mm (3 inches) refractor, was dedicated to hunting comets, the undisputed stars of the 1700s. In his wanderings around the sky, searching for slowly traveling nebulous dots, he ran into, during the course of the years, 110 objects which often resembled comets but never moved.

The famous **Messier catalogue** gathers the brightest clusters, nebulae and galaxies in the sky that can be observed in our latitudes: this is the first source to draw from when beginning to explore deep space.

The objects from Messier's catalogue are identified with an M followed by a number from 1 to 110 and are scattered throughout the sky. For example, the Andromeda galaxy is denoted M31; the great nebula of Orion is M42, and so forth. All sky simulation software will tell us precisely where Messier's objects are. All of them are visible even with good binoculars and therefore become evident little by little as we increase the power (the diameter!) of our telescope.

Instruments of 90mm (3.5 inches) will often let us see shapeless, bright globs, whereas a 200mm (8 inches) Dobson (which could cost less than the refractor!) will let us see the details and characteristics of everything. The images and descriptions of all the objects of Messier's catalogue are gathered on my web site; just follow this address:

http://www.danielegasparri.com/messier/index_messier.html

When we're a little more expert, or if we'd like to enjoy a little challenge, we could dive into the **NGC catalogue** that contains more than 7800 objects from the deep sky, this time including things from the southern hemisphere. Compiled in the second half of the 1800s, when telescopes were much more powerful than Messier's were, it contains all of the objects from the French astronomer's catalogue and many more that he had either not seen or had decided to ignore.

Some, like Perseus' double cluster, are even visible with the naked eye. Almost all of them are within reach of a 200mm (8 inches) telescope, even though some are faint and difficult to track down.

These objects are catalogued with the acronym NGC followed by a number. The double cluster of Perseus, for example, formed by two open clusters, has the double denotation NGC 869-884.

There are also other catalogues, specialized in different classes, but they contain objects that are increasingly difficult to observe.

I'd say it's not too bad for a beginning. To think that if we were to observe two objects per evening, every evening (without taking into consideration the clouds and the Moon), 10 years wouldn't be enough time! If we're disappointed in the planets because there are only four that are interesting, we certainly can't complain now that there aren't enough possibilities!

To each season its objects

Perhaps already by now and certainly with the passing of time we'll understand which class of deep-sky objects we prefer. Mine are the globular clusters because they show a lot of details, but I don't scorn galaxies, although they are much more elusive. Although the sky is very big and every night it offers us a large variety of celestial bodies, it's best to know that there are seasons better indicated for galaxies observation and others for star clusters and nebulae.

The milky strip of the Milky Way is more obvious in the summer and winter. The great amount of stars, gas and dust almost completely obscures the weak light of the other galaxies set behind it, so these seasons are the most appropriate for the nebulae and the star clusters.

Fall, and above all spring are the best periods for observing the galaxies. Between March and June, we'll have the maximum concentration of these flakes thanks to the presence of an agglomerate of the Virgo cluster, an agglomerate of hundreds of galaxies that orbit around each other, many of which are visible with our telescopes at the borders between the Leo and Virgo constellations. In this area of the sky, sweeping casually with our instrument at a low magnification (or even with good binoculars), we'll meet dozens of indistinct flakes. This is how the Universe welcomes us among its most beautiful and imposing creations.

It's impossible for me to talk about all the deep-sky objects that we can observe with our telescope and I'd prefer not to, even if I could, because it's no fun exploring if we already know what's waiting for us. So, I'll limit myself to listing the most beautiful and easiest objects to observe, just to whet your desire to discover more.

Now that we know the sky a little better, I'll no longer distinguish them according to their seasons, but on the basis of the type of object, even inserting a category I haven't talked about yet: double stars.

From now on, everything is in our hands. I don't have much more to teach, because now we can live by our emotions and no one has the right to teach those to anyone. We're at the top of a long stairway that leads almost to the stars; now it's up to each of us, alone, to take the last and most important step to live the Universe the way we like the most.

Double Stars

Double stars are the easiest deep-sky objects to observe because they are generally rather bright.

More than half the stars of our Galaxy aren't isolated but possess at least one companion with which it divides its orbit. The multiple systems are therefore formed by at least two stars that orbit around a common center of mass.

The telescopic observation of the double stars is very rewarding and is easy to conduct with any telescope, which will reveal details and colors very similar to the ones seen in photographs: it's one of the very rare cases in which a photograph shows exactly what a well-trained eye can perceive.

Almost all real double stars appear as a single component to the naked eye and manifest their nature only on the telescope.

Albireo, magnificent double star in the Cygnus constellation. The two components are separated by 34" and have very different colorations.

The sky is rich with highly spectacular multiple systems: the most beautiful is *Albireo* (β *Cygni*), formed by two components – one is orange colored, the other is blue – separated by 34", within reach of every telescope used at magnifications greater than 30 times. The view is truly exciting, thanks to the evident play of colors which testifies to the profound diversity between the two stars.

Another very interesting system is represented by ε (*epsilon*) *Lyrae*, composed, incredibly, by 4 components! The main two are separated by 211" and constitute an excellent test for the naked eye: a perfect eye is able to separate them, with difficulty, without an optical auxiliary. Binoculars and any small telescope show the two components as perfectly separate.

An 80mm (3 inches) refractor or a 114mm (4.5 inches) reflector, used at least at 150X, will let you understand that each of these two components is actual double, with separations of circa 2.5". These constitute an excellent test for the optical qualities of a 60mm telescope (2.4 inches); only if the instrument is perfectly worked will it show the single stars. Tight double stars (those with angles of separation lower than 5") even constitute, therefore, severe tests of the optical quality of our instruments.

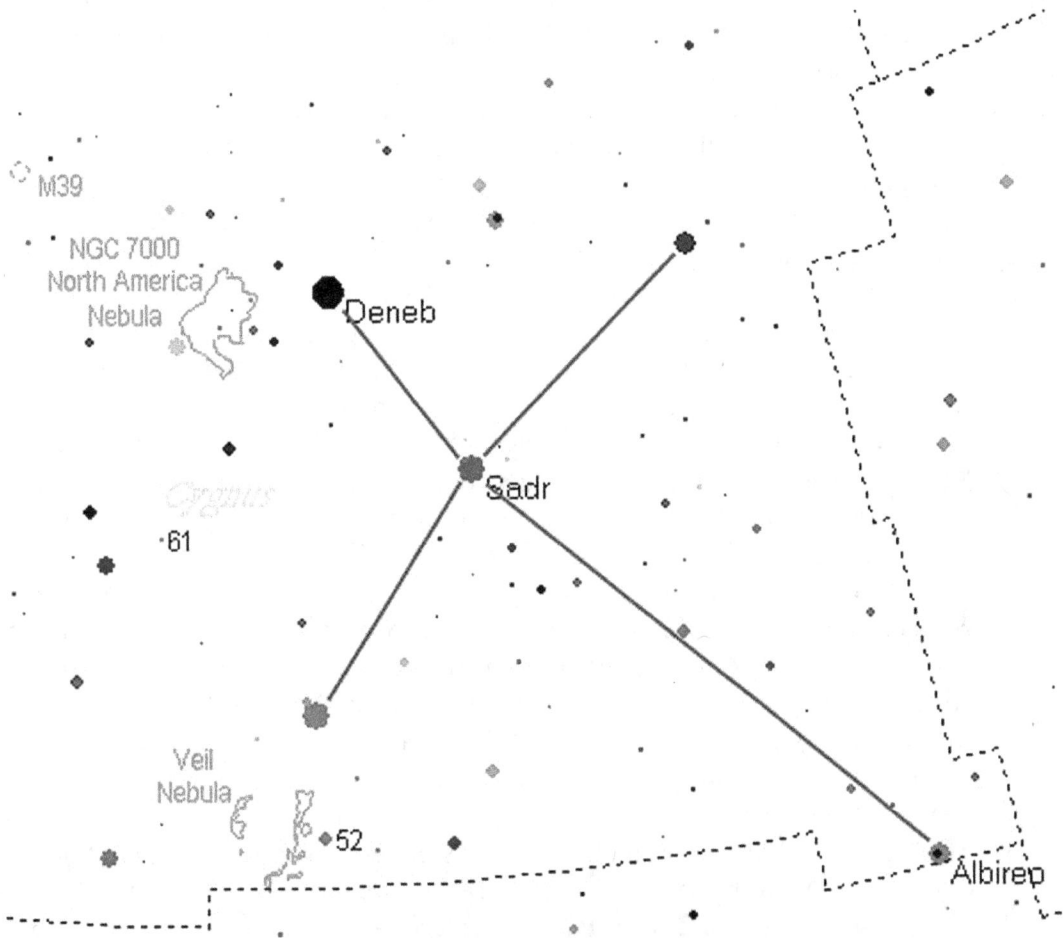

Albireo and the double-double Epsilon Lyrae are among the sky's most beautiful double stars, quite visible in the summer and autumn. Albireo separates with any telescope. The two main components of Epsilon Lyrae can be separated even with the naked eye if we have perfect vision. The four components are observed even with a 60mm (2.4 inches) refractor at a high magnification.

Open clusters

Open clusters are fairly young objects, astronomically speaking. There are very few open clusters older than a billion years, because these objects are destined to dissolve after a few hundred million years.

It's thought that even the Sun was born 4.6 billion years ago as a component of a cluster composed by at least fifty or so stars. Some of these, the more massive ones, exploded ages ago; the others have simply followed a different road, dissolving the initial open cluster. On one hand, this is too bad, because the sky would have been much richer with bright stars, whereas on the other hand, it's a good thing because the eventual explosion of a star like a supernova at a distance less than 150 light years away could potentially cancel every trace of life from the Earth because of the intense emission of gamma rays.

Open clusters are certainly the easiest and brightest objects to observe after the double stars. They are found into the disc of the Milky Way, then the best seasons to observe them are winter and summer. Among the most beautiful are:

- **M45**, or rather, the Pleiades, visible to the naked eye, in the Taurus constellation, in autumn and winter. A spectacular cluster at low magnifications, inferior to 50 times, composed by young, blue-colored stars and a tenuous haze, visible only with instruments of at least 150mm (6 inches) and dark skies;
- **NGC869-884,** or rather the double cluster of Perseus, visible to the naked eye as an indistinct cloud in the constellation of the same name, near the border with Cassiopeia. It is, perhaps, the most beautiful open cluster. Any telescope used at a low magnification (30-50) allows you to observe both of the clusters, composed by at least a hundred stars, some of which show colorations that go from yellow-orange to blue. A 150mm (6 inches) telescope will let you increase the number of faint stars and enter into the heart of the two clusters, filling your eyepiece with a great many gems. Like all open clusters, it loses its awesomeness at magnifications exceeding 100 times, because of its great apparent extension;
- **M44**, called the Praesepe or Beehive cluster, in the Cancer constellation, is easy to sight with the naked eye, to the east of the imposing Leo constellation. It is a less spectacular cluster than the preceding one, but it is interesting because it is rather compact;
- **M11**, called the Wild Duck globular cluster is the most concentrated. Its stars, tightly wrapped in a space equal to half the apparent dimensions of the Full Moon (14'), can be seen by any 80mm (3 inches) instrument. A 150mm (6 inches) telescope will allow you to identify all of the faint components that form this curious object, set in the faint constellation of the Scutum.

The Pleiades (M45) are the brightest and best known open cluster. It's easy to see in the autumn and winter nights in the constellation of Taurus. This drawing shows how they appear through a 150mm (6 inches) telescope at 20X.

The double cluster of Perseus (NGC869-884) is visible to the naked eye as a faint condensation along the winter Milky Way, close to the constellation of Cassiopeia. In this drawing, as it appears with a 100mm (4 inches) telescope.

The beauty of the Pleiades captured by a digital camera. Unfortunately, the human eye isn't sensitive enough to show us this view.

List of some open clusters to observe in the sky

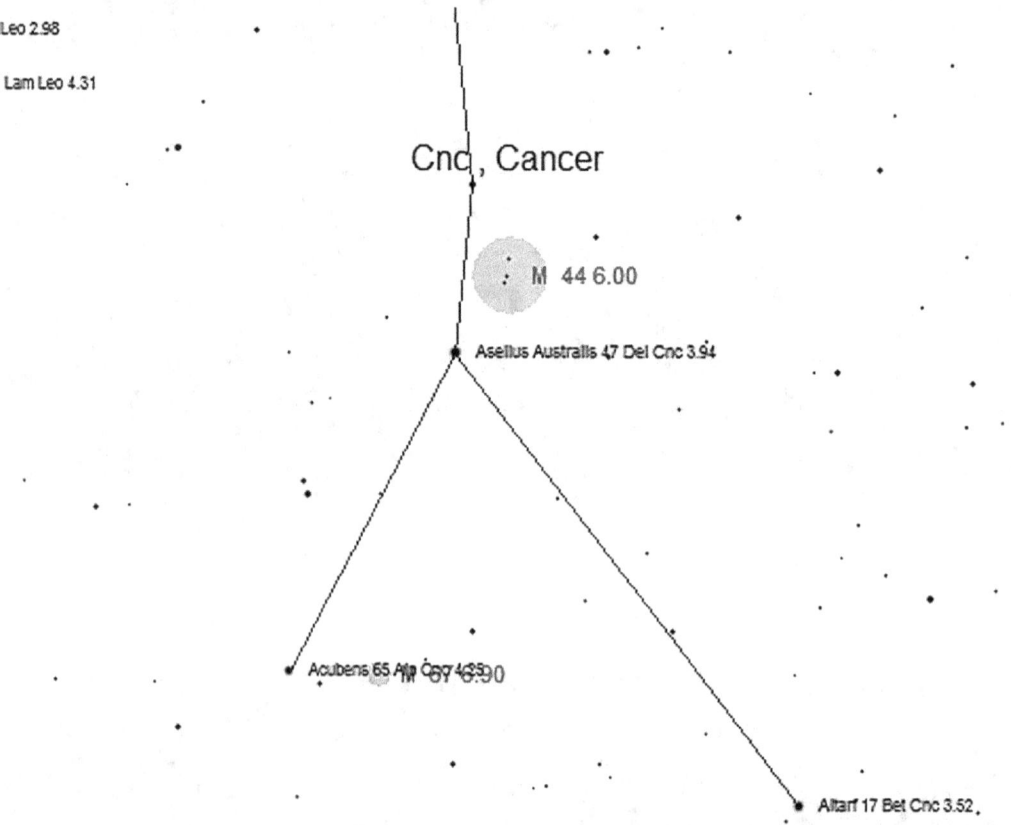

M44 is called the Praesepe cluster and is visible to the naked eye, in the spring and under moderately dark skies. If it's invisible, then the sky you're observing in is too bright and doesn't let you observe faint objects; it's best to aim for the planets. Located in the heart of the Cancer constellation, it's a stupendous object when seen with binoculars and every telescope, as long as low magnifications are used.

Our observations

Sketch Observer's name:

Object and position:		Date:	Time:
Observing site:		Type of telescope:	
Diameter:	Focal:	Eyepiece/magnification:	
Sky darkness:	Seeing:	Transparency:	Moon phase:
Notes and impressions:			

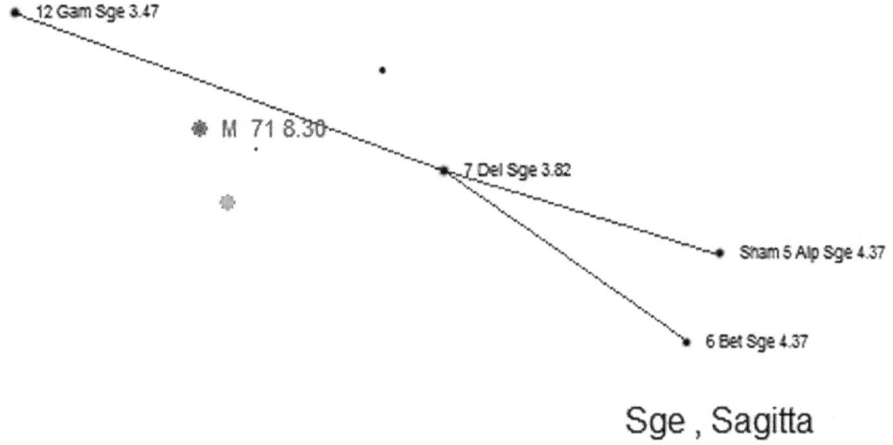

We find M71, a rather faint and concentrated open cluster in the heart of the small constellation of the Arrow (*Sagitta*). Any telescope will show you its image, which will, however, be confused and hazy. The view of the single stars is reserved for instruments of at least 150mm (6 inches), under, naturally, dark skies which are as important, if not more so, than the diameter of our telescope.

Our observations

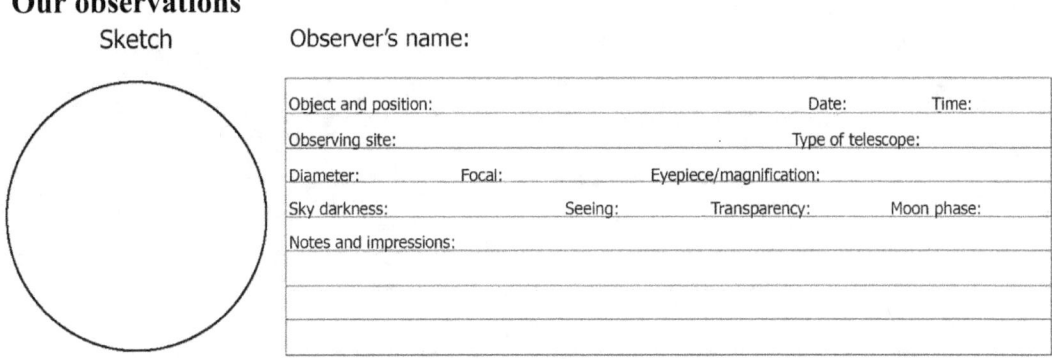

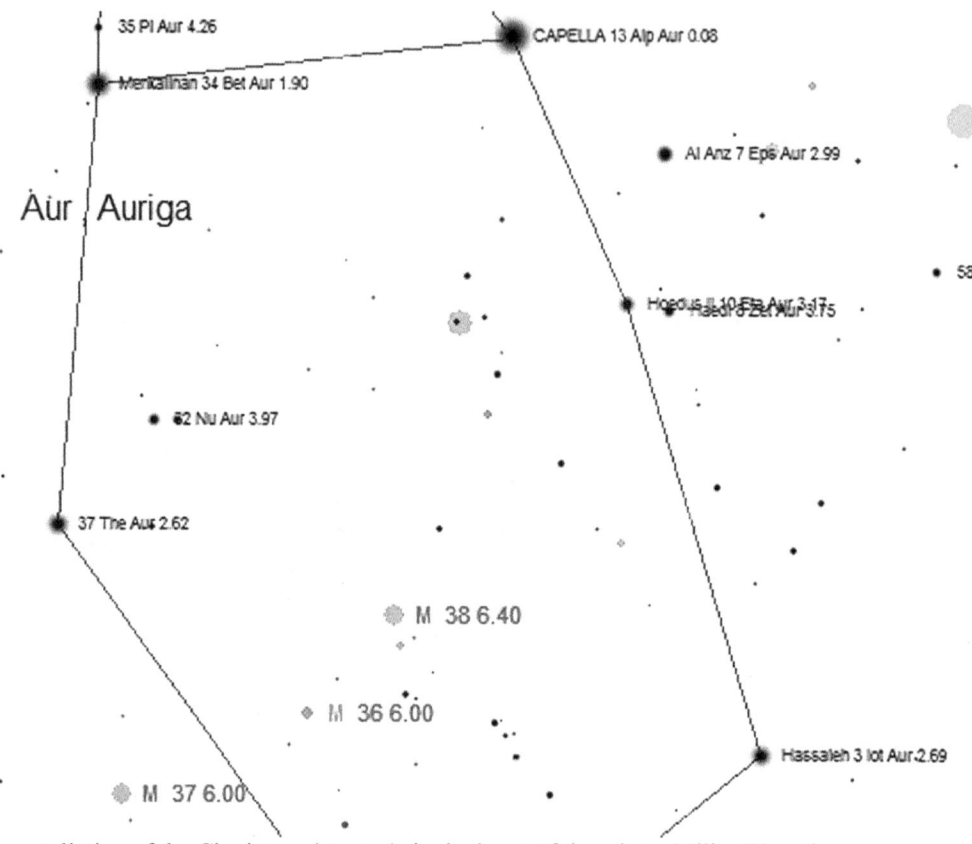

In the constellation of the Charioteer (*Auriga*), in the heart of the winter Milky Way, there are numerous open clusters, some of which are visible even to the naked eye. M36-37-38 are very easy objects to observe with any instrument and truly spectacular. A glance at the heart of this constellation will show you numerous other agglomerates, with different colorings and forms.

Our observations

Sketch   Observer's name:

Object and position:		Date:	Time:
Observing site:		Type of telescope:	
Diameter:	Focal:	Eyepiece/magnification:	
Sky darkness:	Seeing:	Transparency:	Moon phase:
Notes and impressions:			

Let's observe with the telescope

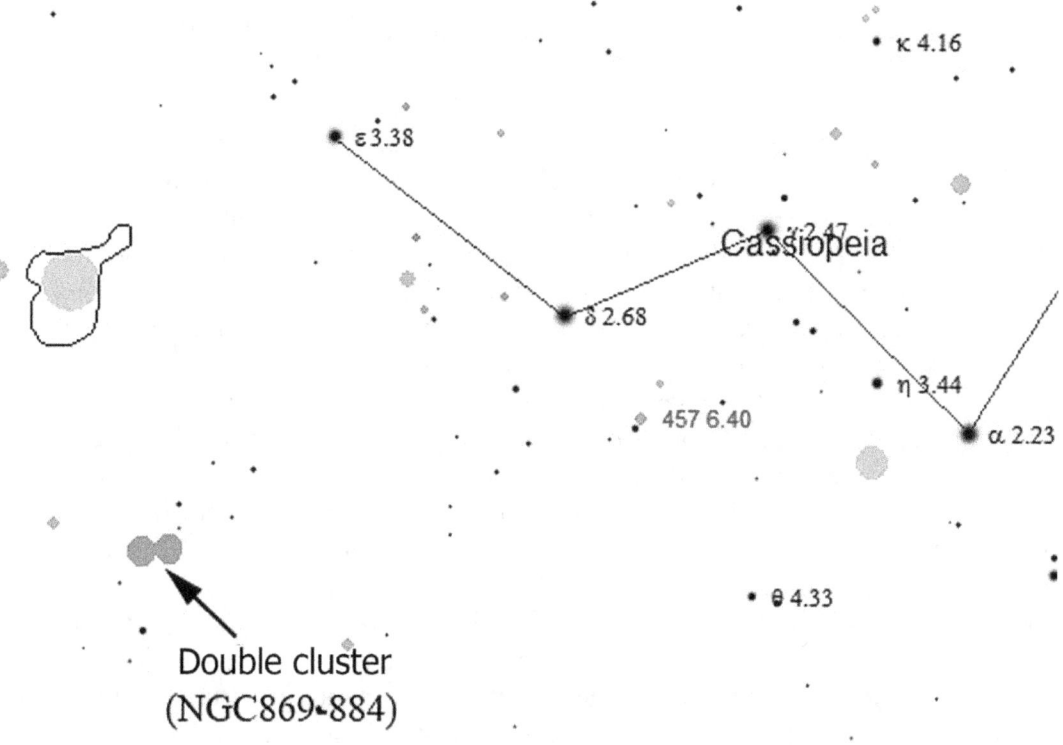

The double cluster of Perseus is formed by two open clusters prospectively close, visible to the naked eye under dark skies. This is, probably, the most beautiful star cluster to observe, full of stars and colors that fill the field of the ocular of any instrument. In contrast to the other objects, viewing open clusters is suggestive with any telescope; they are very much like those in photographs.

Our observations

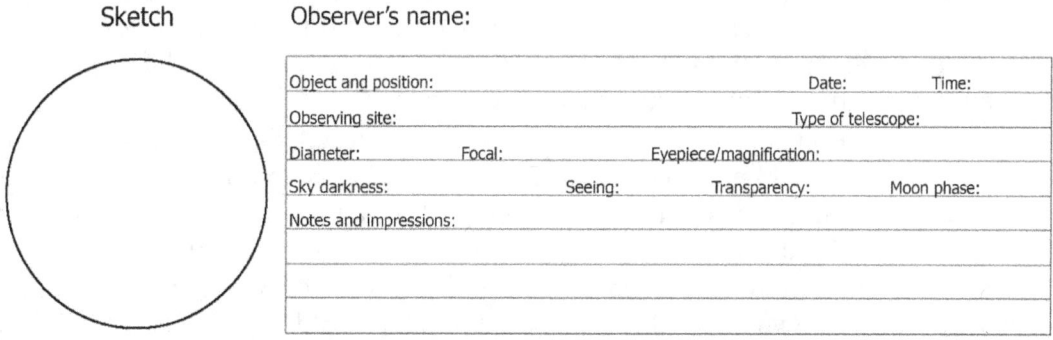

Globular clusters

These are the oldest inhabitants of the Universe and are found in the glow of the galaxies.

Globular clusters are composed by hundreds of thousands of stars grouped into the space of some dozen light years. It's thought that, for all intents and purposes, these can be considered nuclei of primordial galaxies. Almost all of the globular clusters, in fact, are older than the galaxies that host them, most likely a sign that the union of these objects – which populated the young Universe in large numbers more than 10 billion years ago – contributed then in creating those very galaxies. The clusters that we can observe, therefore, are on a par with asteroids regarding the Solar System: fossil relicts of ancient and violent processes of star formation.

They're spectacular with a telescope if it's possible to focus the single stars.

All globulars in the Galaxy can be observed with a 80mm (3 inches) telescope. Some, the largest, can even be seen with modest binoculars or the naked eye if the sky is dark.

Globular clusters are perhaps the objects for which we feel the need for a telescope with an increasingly larger diameter. Instruments from 60-80mm (2.4-3 inches) only show a small condensation without any star structure, very similar to a nebula (and I know this very well!). Telescopes from 100-120mm (4-4.7 inches) begin to show an obvious granularity in the nearest and brightest clusters. The single stars still can't be observed, but it's possible to intuit the object's stellar nature. Instruments of at least 150mm (6 inches) start to resolve the entire structure of the main ones – among them M13 and M22 – when using magnifications of at least 100X. The eyepiece's field teems with tiny but distinct stars that seem to explode in the center of the cluster. All of them are older than 10 billion years; many seem almost as ancient as the Universe. Exciting! The observation of globular clusters with telescopes ranging from 200mm (8 inches) to 250mm (10 inches) is perhaps the most exciting activity of an amateur astronomer: we'll never find other objects where we can observe thousands of stars in such a small space!

Although they're similar, no globular cluster, if carefully observed, is the same as another. A nice challenge is exactly in understanding the differences between the various objects, perhaps with the aid of a pencil and the classical note-pad.

The most beautiful globular clusters to observe are:

- **M13:** the great cluster of Hercules, in the constellation of the same name, passes right overhead on the hot summer nights. Its integrated magnitude (meaning the total) is 5.9 and it constitutes an excellent test for evaluating the worth of an observation site. If you can see it with your unaided eye, at least with averted vision, without the help of any instrument, then the sky where you've observed it is good quality and can give you a good satisfaction. From the world's best skies, M13 is easy to see, even with the naked eye, even looking directly. It's already bright and evident in small telescope finders and instruments with a modest diameter, like 60-70mm (2.5 inches) refractors or 90-114mm (4.5 inches) reflectors. Its brightest stars have an 11.5

magnitude and a telescope of at least 120mm (4.7 inches) is necessary to begin unveiling its stellar nature. A 200-250mm (8-10 inches) telescope, used at a magnification of 100X, fills the eyepiece's field with thousands of very faint, little stars: one of the most beautiful and exciting views to observe.

- **M22** is brighter than M13, but it's found in the constellation of Sagittarius and, therefore, always low on the horizon. If you have a dark sky available near the southern horizon, you can observe and resolve it, even starting at 114mm (4.5 inches).

The apparent dimensions, similar to those of the Moon make it the objective of widefield eyepieces, with which you'll have the impression of flying over it. With 150mm (6 inches) telescopes and magnifications of 100X it appears almost entirely resolved in the single components.

- **M4** is less spectacular but easier to resolve because it's not as dense as the others. It's located in the constellation of the Scorpion, near the bright *Antares*. Very easy to identify, it's best when observed at modest magnifications and with telescopes of at least 150mm (6 inches), although it's visible with all optic instruments.

There are many other beautiful globular clusters to observe, among which M92, M2, M5 and M15. Unfortunately the brightest and most beautiful are found in the southern hemisphere and can't be seen from our latitudes. *Omega Centauri*, which shines like a magnitude 4 star, and 47 *Tucanae*: they are the great gems of the sky.

M3 is a globular cluster in the Canes Venatici constellation. In this drawing, as it appears through the ocular of a 200-250mm (8-10 inches) telescope, observing at about 100X.

M5, in the Snake, is very dense and difficult to resolve up to the center with telescopes under 200mm (8 inches). In this drawing, as it appears with a 250mm (10 inches) telescope.

List of some globular clusters to observe in the sky

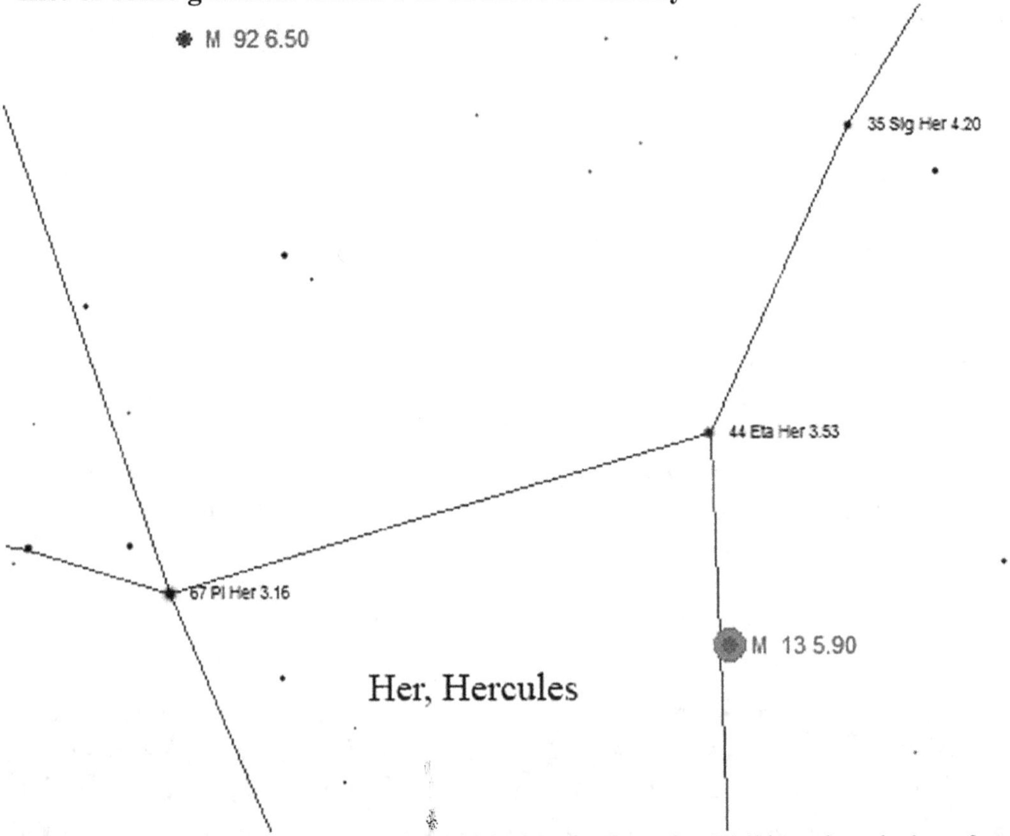

✹ M 92 6.50

· 35 Sig Her 4.20

• 44 Eta Her 3.53

• 67 Pi Her 3.16

⊙ M 13 5.90

Her, Hercules

The globular cluster M13 is the most beautiful of the northern sky. Visible to the naked eye from dark skies, it's very easy to point. Evident with any instrument, it begins to show its stars with a 120mm (4.7 inches) telescope, used at a higher than 100 magnifications. M92 is its slightly smaller "twin brother". Telescopes of at least 150mm (6 inches) are necessary for resolving the stars in globular clusters.

Our observations

Sketch   Observer's name:

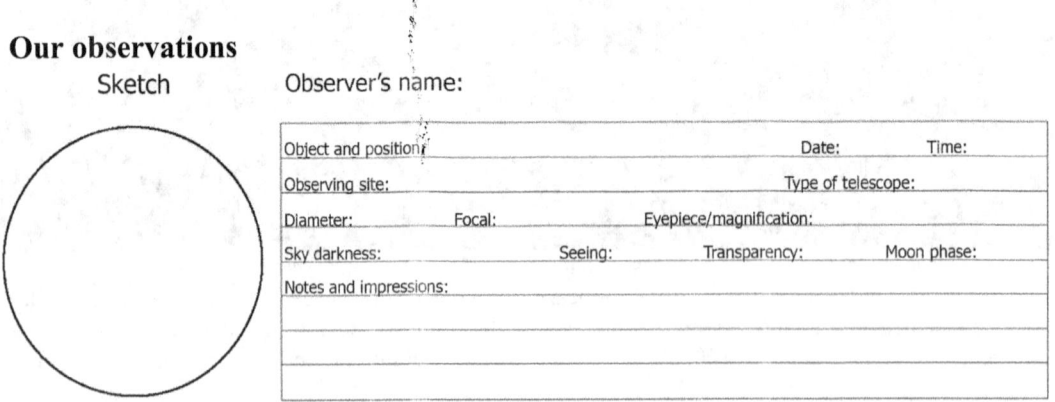

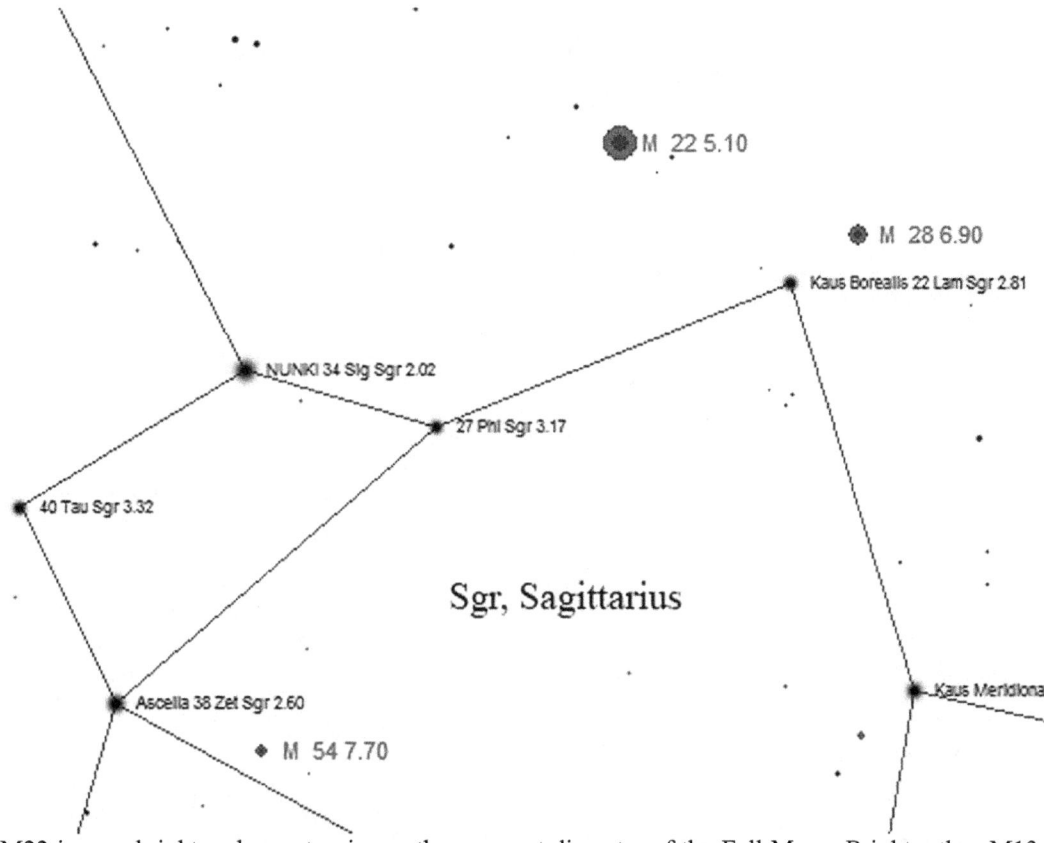

M22 is very bright and as extensive as the apparent diameter of the Full Moon. Brighter than M13, it's penalized by its low height over the horizon (for observers at mid latitudes north). From sites in the mountains, with clear skies all the way to the horizon, it's visible even with the naked eye. Easy to observe with any instrument, it shows its stars, starting from magnitude 11.5, with telescopes of at least 120mm (4.7 inches). A 200mm (8 inches) instrument shows it splendidly resolved.

Our observations

Sketch   Observer's name:

Object and position:			Date:	Time:
Observing site:			Type of telescope:	
Diameter:	Focal:		Eyepiece/magnification:	
Sky darkness:		Seeing:	Transparency:	Moon phase:
Notes and impressions:				

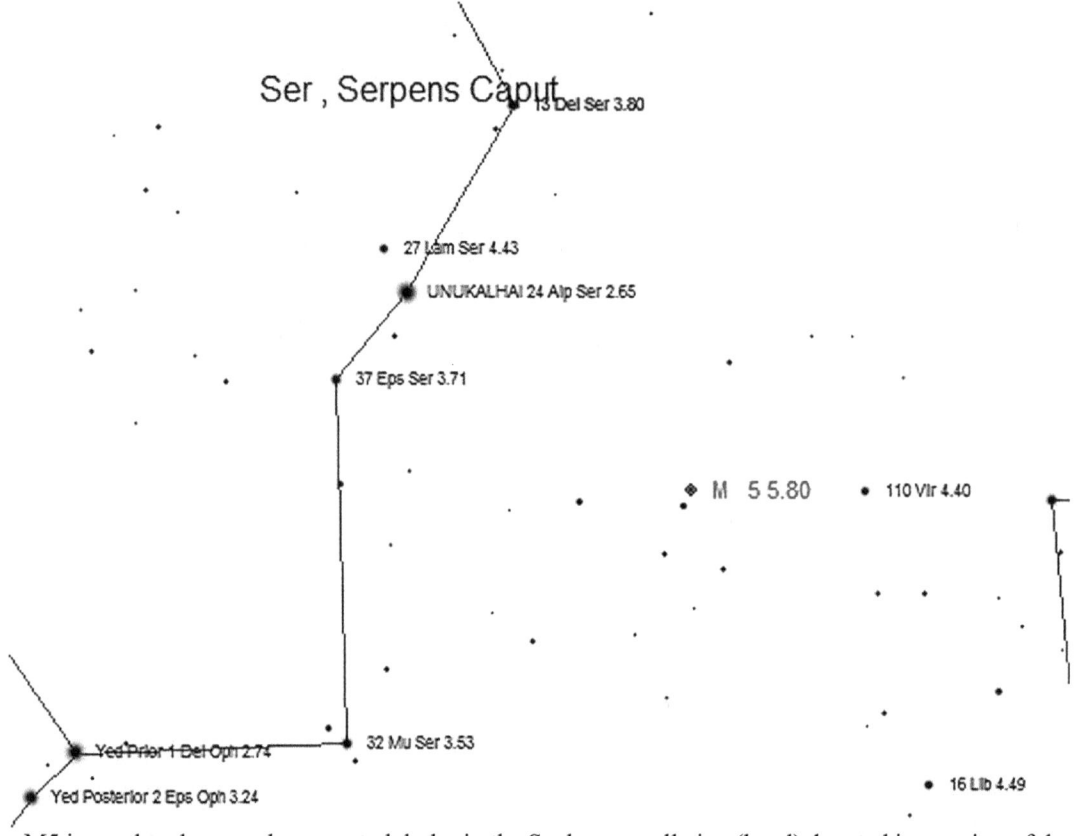

M5 is a rather dense and compact globular in the Snake constellation (head), located in a region of the sky that lacks bright stars. Although visible through telescope finders, it's not immediately traceable. Instruments from 200mm (8 inches), with magnifications of at least 150 times, are needed to resolve its structure.

Our observations

Sketch   Observer's name:

Object and position:		Date:	Time:
Observing site:		Type of telescope:	
Diameter:	Focal:	Eyepiece/magnification:	
Sky darkness:	Seeing:	Transparency:	Moon phase:
Notes and impressions:			

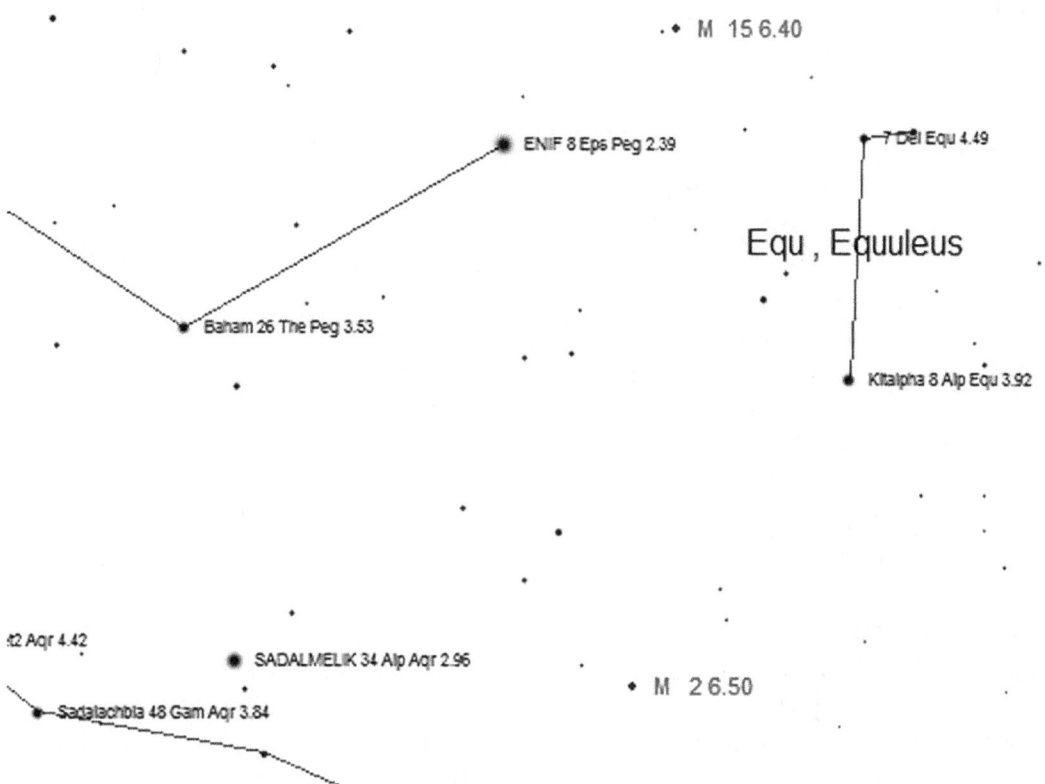

M2 and M15 are globulars observable in the autumn evenings; they're similar in brightness, but much different in shape and density. M2, in the Aquarius constellation, is spherical, and not very dense, so instruments from 150mm (6 inches), under very dark skies, show a good definition of its structure. M15 is much denser, so much so that instruments from 200-250mm (8-10 inches) are required for resolving it all the way to the center.

Our observations

Sketch   Observer's name:

Object and position:			Date:	Time:
Observing site:			Type of telescope:	
Diameter:	Focal:	Eyepiece/magnification:		
Sky darkness:		Seeing:	Transparency:	Moon phase:
Notes and impressions:				

Nebulae

The nebulae are perhaps the most interesting objects for beginners because they are different from everything we're used to seeing; and yet, at times they are miserly with their details and difficult to observe profitably.

Nebulae are immense stretches of extremely rarefied gas.

With this term, we are referring to a gas with a much lower density than the highest vacuum that can possibly be created here on the earth. The typical density of nebulae is about 1000 particles to every cubic centimeter of space. Just think that the Earth's atmosphere at sea level contains something like 10^{19} molecules (1 followed by 19 zeroes, an impossible number to pronounce!).

A nebula represents both the stage preceding the life of all stars and its final act. All stars are born from nebulae and end their existence as nebulae, enriching interstellar space with new material, used in turn by the next generation.

From this point of view, we can think of stars as living beings: they're born, they develop and, at the end of their life, they die, expelling the material that they are composed of which will not, however, be lost but will be a fundamental part of the birth of other stars.

There are different types of nebulae. Those that give birth to stars are very extensive, at least ten light years and, depending on the various evolutional phases, take on different properties and names.

When there are still no stars in the nebula, it's extremely cold (approximately -436°F) and rather dense; however, it is completely invisible: these are dark nebulae.

When the first stars begin to form – through the compression of a part of the gas – inside this cloud, the nebula is heated up by the radiation and if it exceeds a temperature of 18,032°F, it emits its own light, of a light reddish coloration. This type of nebula is called emission nebula.

When the stars' light isn't sufficient for heating the gas, or else the hotter and more massive ones are already extinct, the nebula diffuses and reflects the light of the stars contained or observed behind its line of sight and, similar to a fog bank illuminated by a lighthouse, is made visible. These nebulae are called reflection nebulae.

When a star exhausts the fuel inside it (mainly hydrogen), it heads toward the end of its existence. Some, depending on the amount of material it contains, can burn helium and other elements, passing through various phases, called red giant or red super-giant (only for the most massive).

Sooner or later, in any case, the star's life is doomed. If its mass is 8 times less than that of the Sun, the red giant, and has finished burning its helium, it begins to gradually expel the most external levels, in a process that, in a few thousand years, will give life to a splendid planetary nebula.

If the star's mass is greater than 8 times that of the Sun, its end will be much more violent. After having consumed all of the fuel at its center (until it forms an iron nucleus), the star suddenly collapses in on itself. The immense shockwave provokes an enormous explosion, called supernova, which destroys a large part of the stellar structure and hurls the matter it is made from into space, at a speed of tens of thousands of miles per second.

During this explosion, which typically lasts around a month, the energy levels are so high that, due to nuclear fusion, all elements that are heavier than iron – among which gold, silver, platinum and many other metals – are produced. In fact, all elements present in the Universe and, therefore, also present on the Earth, with an atomic weight higher than that of iron, have been produced during these very violent phases of the death of a massive star. If you have gold jewelry at hand, now you know that it comes from the explosion of a star, in who knows what part of the Galaxy, a very long time before the formation of the earth and the Sun (4.6 billion years ago).

The energy released during every second of a supernova's explosion is frighteningly high, greater than that emitted by the entire galaxy in which it explodes. In fact, the *supernovae* are visible even from a distance of hundreds of millions of lightyears away.

With the passage of time, when the initial energy has faded, the signs of this explosion are manifested with the appearance of a nebula, called supernova remnant, which is nothing more than the pieces of star that were hurled into space.

Planetary nebulae and supernovae remnants, originating from the end of a single star, are reduced in size, decidedly inferior compared to the great expanses of gas from which they are born and develop. Besides, these objects have a relatively short life, rarely longer than 100,000 years.

All bright nebulae (and therefore, with the exception of the dark ones) can be spectacular with a telescope, above all now that the properties and mechanisms associated with them are known. The important thing is to forget the photographic views that you may have already seen in books and magazines: the nebulae will never show themselves to you in those colors and extensions.

They are objects that are very, very different from how they appear in every photo: there is no instrument, no matter how large and powerful, that will show up like a long-exposure photograph.

The most advisable nebulae, full of details, are the small planetary nebulae, among which we can mention M57, the famous Ring Nebula in the Lyra constellation and M27, in the small Fox constellation (*Vulpecula*).

The planetary nebulae are angular and therefore a magnification of at least 100 times is required to be able to admire them satisfactorily. Fortunately, they are also intrinsically brighter than the others, and are, therefore, easier to observe. The sky is full of these small objects with the most disparate shapes.

Emission nebulae are much more extensive, full of shadings but also much, much fainter.

The emission nebula par excellence is surely the great Orion Nebula (M42), in the heart of the constellation of the same name; it's easy to identify, even with the naked eye, as a blurry star.

If observed in a dark sky, M42 is the most beautiful nebula of all. A small 80-100mm (3-4 inches) telescope allows the view of even the 4 stars present at its center, the famous Trapezium. Using averted vision, one has the impression of seeing the image of a bird flying freely in the sky.

Instruments from 200-250mm (8-10 inches) present a dreamy image. Although colorless, except for a soft blue-green tint in the central areas, the nebula actually appears similar to photographs taken of it. The shades of gas are evident, just like the different intensities and wefts that run along this immense expanse of gas. No other emission nebula is this spectacular, even though there are other interesting ones, like M8, the Laguna Nebula in the Sagittarius constellation, which is easily visible with the naked eye; M20, called Trifid, 1.4° north of the Laguna; then, M16, called the Eagle Nebula, and its neighbor (prospectively speaking), the Omega Nebula. Each of these nebulae has a particular shape and shade, which only training, averted vision and a dark sky will allow you to admire them in all of their splendor.

All emission nebulae are more extensive than the Moon's diameter; a profitable observation leads, therefore, to low magnifications (around 50X) and possibly with wide-field eyepieces.

Reflector nebulae are dim and difficult to observe.

Also worth mentioning are: the slight haziness around the Pleiades cluster, visible with instruments from at least 150mm; the northern part of the Trifid Nebula, easy prey of every instrument and M78, reflector nebula in the Orion constellation.

Supernova remnants are very rare: only M1 can be seen by every instrument, but it doesn't give any particular emotions, if not with telescopes greater than 200mm (8 inches) and expert eyes.

Dark nebulae are best observed with the naked eye and with binoculars, looking along the galactic disc, especially in the summer.

The famous Horsehead Nebula, perhaps the most famous of this type, is observed only with telescopes greater than 250mm (10 inches).

A dark sky is more essential than ever and determines the nebula's aspect as well as the visibility of its lighter parts which fade softly into the deep sky. Naturally, the Moon's presence over the horizon disturbs the view and more than just a little; so much so, in fact, that, the total absence of even a hint of a crescent Moon in the sky is necessary in order to better observe these objects and the galaxies. If this isn't the case, then it's better to look for something else: planets, open clusters or even the Moon itself.

More demanding and expert observers can improve their view of emission, planetary and supernova remnant nebulae by using nebular filters or a more selective OIII filter, centered on the green emission of two-time ionized oxygen, typical of these expanses of hot gas. Dark skies, telescopes of at least 200mm (8 inches) and an excellent OIII filter are the ideal protagonists of showing shades and shapes that are very difficult to forget.

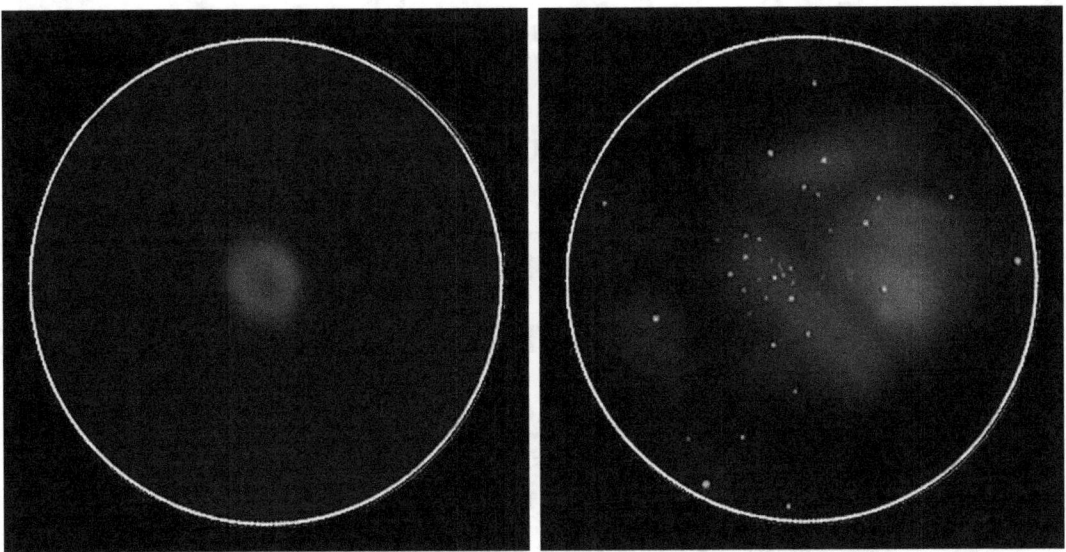

Appearance of a few nebulae through the eyepiece of a telescope of about 150-200mm (6-8 inches), under a very dark sky. **At left:** the small planetary nebula M57 in the *Lyra* constellation. **At right:** the great emission cloud called Laguna (M8) in the heart of the summer Milky Way.

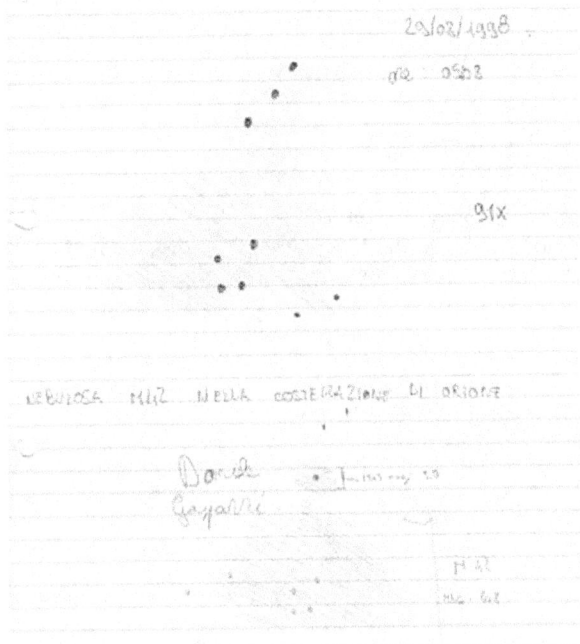

The early drawings and observations of a young amateur astronomer (me!) discovering the Universe. The great Orion nebula, with the central trapezius surrounded by a slightly blue-green haze.

List of some nebulae to observe in the sky

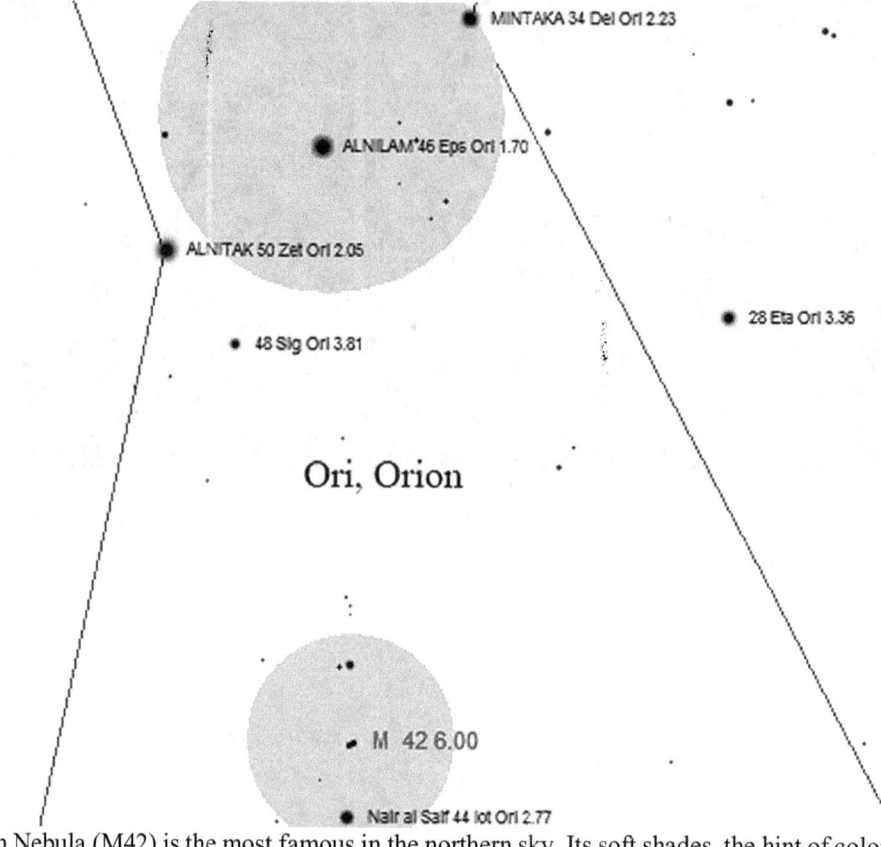

The great Orion Nebula (M42) is the most famous in the northern sky. Its soft shades, the hint of color at the center, where the open cluster called the Trapezius is found, make this object one of the most beautiful.

Our observations

Sketch   Observer's name:

Object and position:		Date:	Time:
Observing site:		Type of telescope:	
Diameter:	Focal:	Eyepiece/magnification:	
Sky darkness:	Seeing:	Transparency:	Moon phase:
Notes and impressions:			

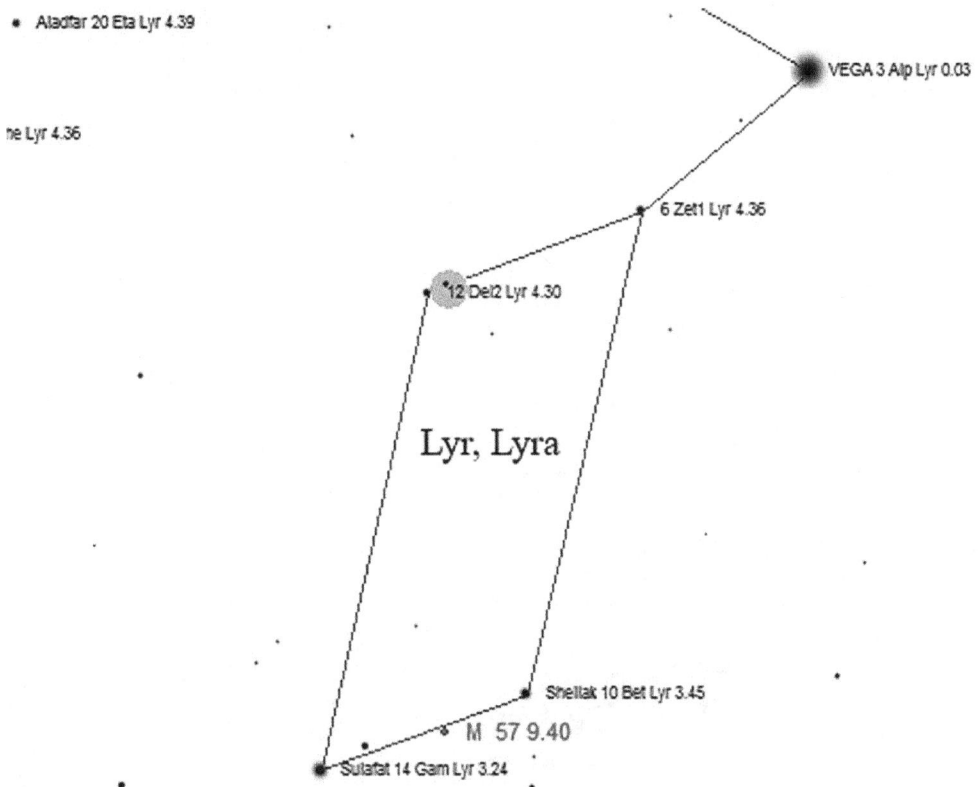

M57 is the sky's most famous planetary nebula. Nicknamed the Ring Nebula because of its shape, it's visible, although with difficulty, with instruments of at least 100mm (4 inches). Like any planetary nebula, it has apparently modest dimensions but a high superficial brightness, making it visible with any instrument.

Our observations

Sketch   Observer's name:

Object and position:		Date:	Time:
Observing site:		Type of telescope:	
Diameter:	Focal:	Eyepiece/magnification:	
Sky darkness:	Seeing:	Transparency:	Moon phase:
Notes and impressions:			

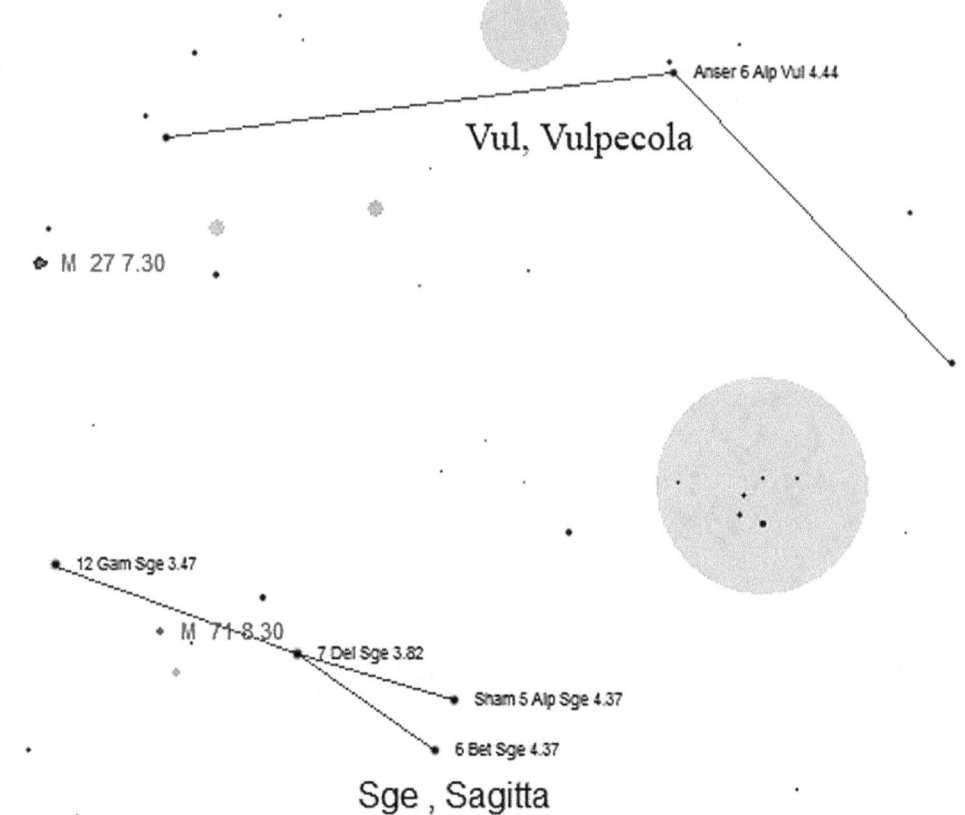

M27 is another planetary nebula, in the faint constellation Vulpecula. Easy to observe even with binoculars, it's at its best through a telescope. Its particular shape and the faint shades it presents are within reach of telescopes from 200mm (8 inches), helped, possibly, with a nebular filter. The OIII filters are excellent for planetary and emission nebulae.

Our observations

Sketch Observer's name:

Object and position:		Date:	Time:
Observing site:		Type of telescope:	
Diameter:	Focal:	Eyepiece/magnification:	
Sky darkness:	Seeing:	Transparency:	Moon phase:
Notes and impressions:			

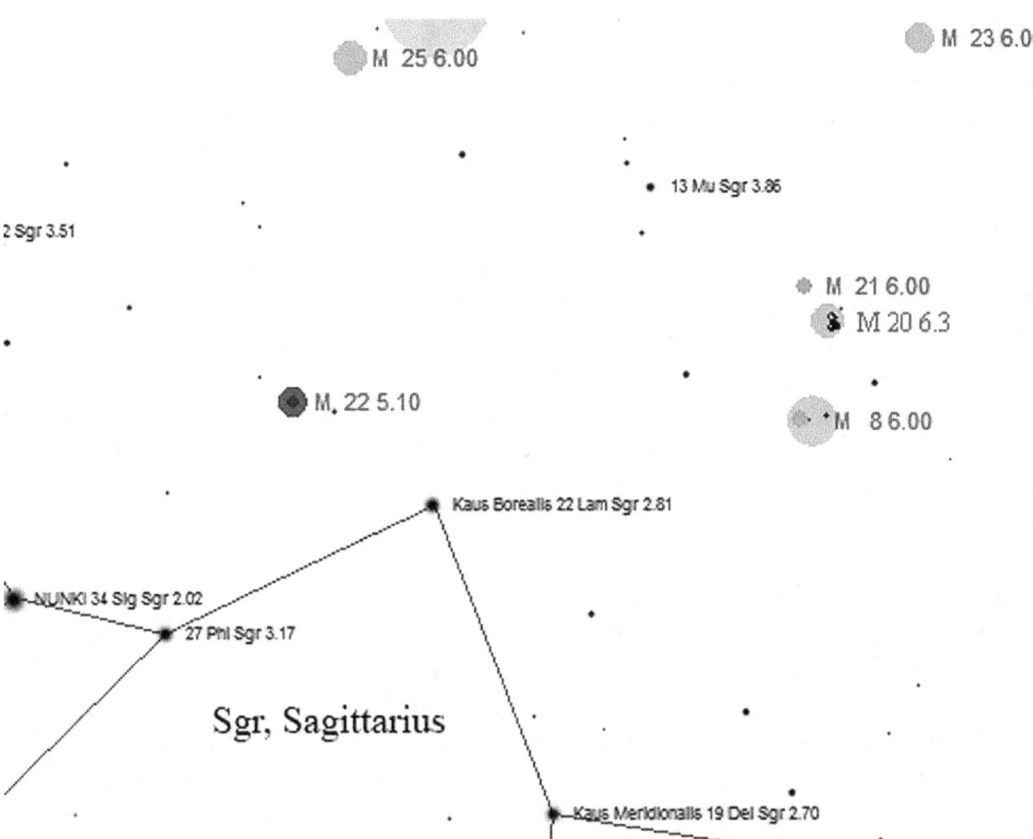

During the summer nights, to the south, we the find the Sagittarius Constellation, placed over the center of our Galaxy. The Laguna Nebula (M8) as a small, indistinct cloud. M24, called the Trifid Nebula, is found 1.4° to the north, visible even with binoculars. They are magnificent targets with 150mm (6 inches) or greater telescopes.

Our observations

Sketch   Observer's name:

Object and position:		Date:	Time:
Observing site:		Type of telescope:	
Diameter:	Focal:	Eyepiece/magnification:	
Sky darkness:	Seeing:	Transparency:	Moon phase:
Notes and impressions:			

Galaxies

All of the objects that we have seen up to now belong to our Galaxy, the Milky Way.

It is thought that at least 500 billion galaxies exist in the Universe, containing hundreds of billions of stars, millions of nebulae, thousands of star clusters and, with all probability, billions of other planets. The Universe isn't limited to what we can see with the naked eye; it is much, much larger than what we may think and humanly imagine. Just think that, despite the 500 billion galaxies, more than 90% of the Universe is simply empty (or almost): can you imagine a volume so immense that such a large part can remain empty, despite the presence of 500 billion galaxies, each one formed by hundreds of thousands of Suns?

The fascination of being able to observe other galaxies, to push ourselves almost to the edge of the edges of the Universe represents one of the strongest sensations that astronomical observation can give us. Unfortunately, observation of the galaxies is perhaps the most difficult to conduct and the stingiest with details.

In fact, the galaxies are extremely faint objects, often having reduced dimensions, basically because of their extreme distance from us.

The great Andromeda galaxy is the closest one to us, barely (so to speak!) 2.3 -2.4 million light years away). Beautiful and exciting, especially the first times, using binoculars and telescopes with low magnifications (20-30X), it gives nothing else: the stars, the clouds of gas and the spiral arms that are so beautiful in the photographs, aren't visible with any instrument. It's only possible to discern any of the heterogeneities along its disc – caused by the abundant presence of cold gas – with telescopes from at least 200mm (8 inches).

There are many other galaxies in the sky, among which:

- **M33**, in the Triangle, is the second closest galaxy, 2.5 million lightyears away. It shines with a magnitude of 5.7 and can be observed with the naked eye in very dark skies. It shows up almost transparent with a telescope, so much so that it's difficult to view details using instruments with less than 250mm (10 inches).

- **M51:** the famous Whirlpool Galaxy is a spiral in the constellation of the Hunting Dogs, slightly below the handle of the Big Dipper. It's probably the most famous thanks to the way it shows its arms. Don't be deceived, though: instruments with diameters lower than 250mm (10 inches) will show you only two balls of light, the main galaxy and the small companion with which it interacts. Larger instruments, beginning with 100X magnifications, will reveal – faintly – its spiral arms.

- **M81** is another spiral galaxy in the Ursa Major (Big Dipper) constellation. Easy to sight with any instrument, it has apparently considerable dimensions and is evident with a magnification of about 30-50 times. There's no hope of observing its beautiful spiral arms if not using a telescope with a diameter of 20 inches and a very dark sky.

- **M82** is a small irregular galaxy prospectively close to M81. Since it is compact and bright, it's very easy to observe. It's probably one of the few galaxies that proportionally show their details by increasing the diameter of the telescope. An instrument from 80-100 at low magnifications shows it as evident and elongated; it's no coincidence that it's nicknamed the cigar galaxy. A 150mm (6 inches) telescope used with at least 100 magnifications allows the observation of some irregularities in the disc, lined by enormous dark nebulae. A 250mm (10 inches) telescope returns a beautiful view, rich with nuances and details.
- **NGC253** is not well-known but is also one of the few galaxies that show details. It's a spiral, almost showing its profile and, therefore, very elongated. At an integrated magnitude 7, it's easy to see even with 20x80 binoculars, presenting itself as a thin, undefined line. A 150mm (6 inches) instrument is enough to show irregularities in the disc: this is also lined by enormous amounts of cold gas, and, therefore, opaque. We can push ourselves further with magnifications greater than 100 times and try to extract every detail. In fact, we are watching a galaxy that is very similar to our own; it's truly very exciting. A 250mm (10 inches) instrument allows us to have a very detailed view. NGC253 remains one of my favorite galaxies.
- **The Virgo Cluster** is a zone between the Leo and Virgo constellations, rich with galaxies, all linked gravitationally, as though it were an immense open cluster. Jumping in with your own instrument, best if it's at least 150mm (6 inches), with a very wide-field eyepiece and a low magnification, we can observe dozens and dozens of little cottony balls, mainly elliptical galaxies, in a space of around ten degrees. Stop a moment and think, letting yourself be carried away by the greatest gift a human being has: the capacity of traveling with his mind. These small fluffy balls of light are actually immense cosmic agglomerates containing billions of stars that are in turn hundreds of times larger than our Earth. In a space in the sky of just a few degrees you can count dozens of galaxies: do you get an idea of the immensity of the Universe and at, the same time, of the harmony and perfection of the mechanisms that rule it? These galaxies orbit around each other at a speed of several thousand miles per second; every star, each one of the total thousands of billions, follows well-defined rules; nothing is left to chance, no matter how complex and big it is. Terribly fascinating, right?

The sky is full of other galaxies, some easy to observe even with modest diameters, but almost all of them, unfortunately, detail-poor. The enjoyment in observing the galaxies, especially with small telescopes, is in intuiting the shape and in knowing that that small fluffy ball of light contains hundreds of billions of stars, thousands of nebulae and millions of planets. And who knows if there is maybe at least one inhabited by someone who, in this exact moment, is looking with his telescope toward that fluffy ball of light called the Milky Way?

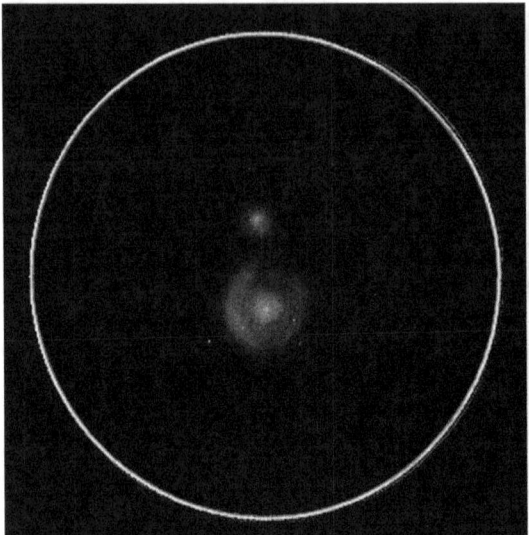 

M51, in the Hunting Dogs, through the eyepiece of a 250mm (10 inches) telescope, from a very dark sky, begins to show its faint swirling arms. In order to show their details, galaxies require telescopes of at least 200mm (8 inches).

NGC 4565, beautiful spiral galaxy, viewed edge-on, in the Virgo constellation, seen with a 200-250mm (8-10 inches) telescope.

Observing the arms of spiral galaxies

Galaxies are the deep-sky objects that I prefer observing; the reason is in the search for a perfect mix of beauty, power, elegance and distance.

However, not all galaxies are easy and exciting to observe.

The most interesting class (and fortunately the most numerous) is represented by the spiral galaxy, splendid cosmic designs shaped by a thin disk that shows the unmistakable swirling arms composed of stars and hot gases. When these objects are observed sideways they appear to have a funny shape, similar to a faded space ship, with an evident central bulge identified by the nucleus.

One of my main challenges has been the direct observation of the arms of those spirals that appear almost perfectly when viewed straight on. I remember that when I was little, I would be completely entranced while observing splendid images that portrayed these immense cosmic spinning wheels in all their beauty and majesty.

Unfortunately, I soon understood (as you will also understand shortly) that the difference between visual observation and a photograph is maximal for spiral galaxies.

Whereas a telephoto lens with a 5centimeter (about 2 inches) diameter is able to show the spiral structure of the brightest galaxies, a 25centimeter (10 inches) telescope is not enough to show them to our poor eye.

Despite the extreme difficulty, I always dreamed of being able to directly see, with my own eyes, these wonderful arms where billions of stars – and probably just as many planets – find their home.

In the end, after many attempts (and a lot of telescopes), I was able to observe, faint but distinct, the spiral arms of M51 which, as we have seen, is the easiest galaxy from this point of view.

The view of an extragalactic object that finally showed itself, even though vaguely, for what it really was and not as one of the many faint and indistinct fuzzy balls, represented for me one of the most sublime moments that observing the sky for all these years has ever offered.

Unfortunately, the observation of these faint details is reserved for very dark skies and telescopes from about 300mm (12 inches). If you don't own one and have no intention of buying it (it's better that way: knowing how to manage a 300mm (12 inches) instrument requires a certain amount of experience), you have yet another reason to observe with telescopes of other amateur astronomers during star parties and observational evenings.

Believe me, the view will be unforgettable.

After M51, M101 is the frontal-view spiral that most easily shows its faint arms. In this drawing, as it appears with a telescope with a 400mm (16 inches) diameter.

List of a few galaxies to observe in the sky

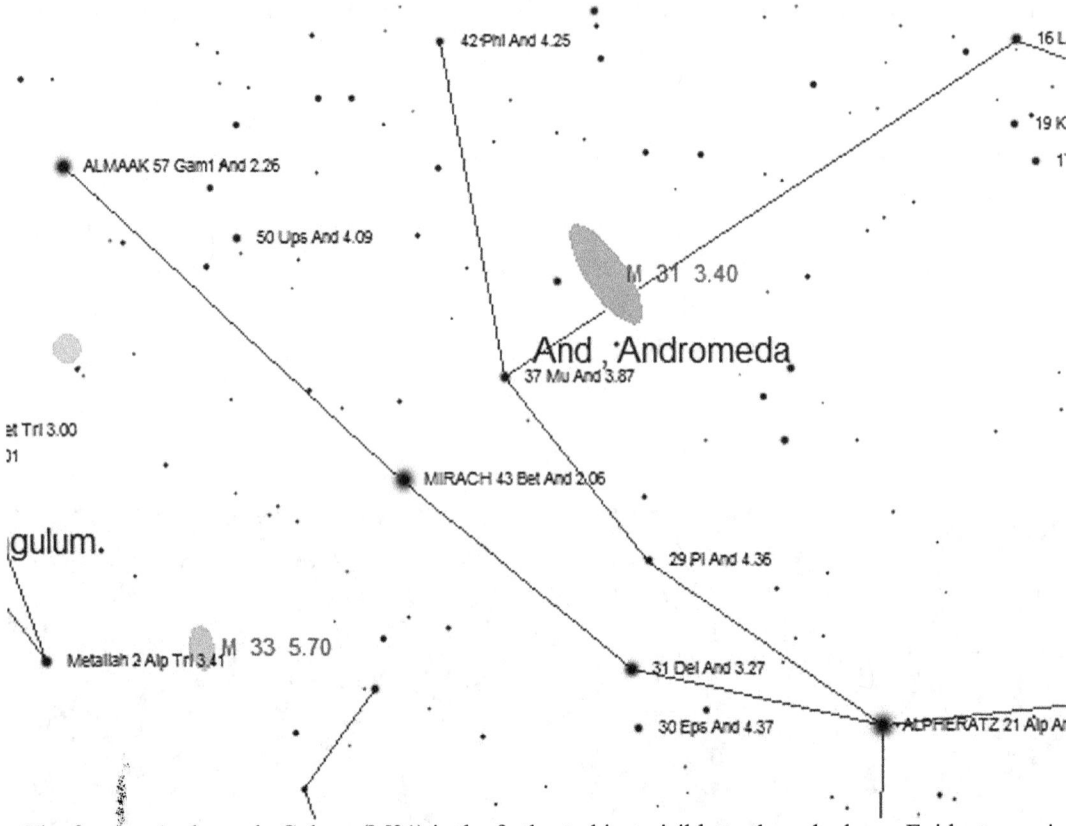

The famous Andromeda Galaxy (M31) is the farthest object visible to the naked eye. Evident even in skies that aren't very dark, it shows almost no details with any instrument, although greater diameters show an increasing contrast. It is probably the most diffused object which shows fewer details when observed despite the instrument's power.

Our observations

Sketch   Observer's name:

Object and position:		Date:	Time:
Observing site:		Type of telescope:	
Diameter:	Focal:	Eyepiece/magnification:	
Sky darkness:	Seeing:	Transparency:	Moon phase:
Notes and impressions:			

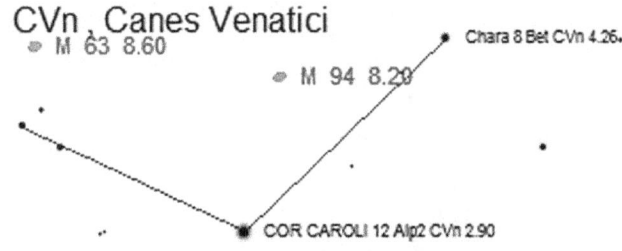

There are many bright galaxies in the Hunting Dogs constellation, among them M51, a very beautiful spiral galaxy, the arms of which are the easiest in absolute to observe with a telescope. Despite this, telescopes from 250mm (10 inches) and very dark skies are necessary for observing them; this says everything about the fundamental role played by an instrument's diameter in deep-sky observations.

M106, M63 and M94 are other interesting spirals.

Our observations

Sketch   Observer's name:

Object and position:			Date:	Time:
Observing site:			Type of telescope:	
Diameter:	Focal:		Eyepiece/magnification:	
Sky darkness:		Seeing:	Transparency:	Moon phase:
Notes and impressions:				

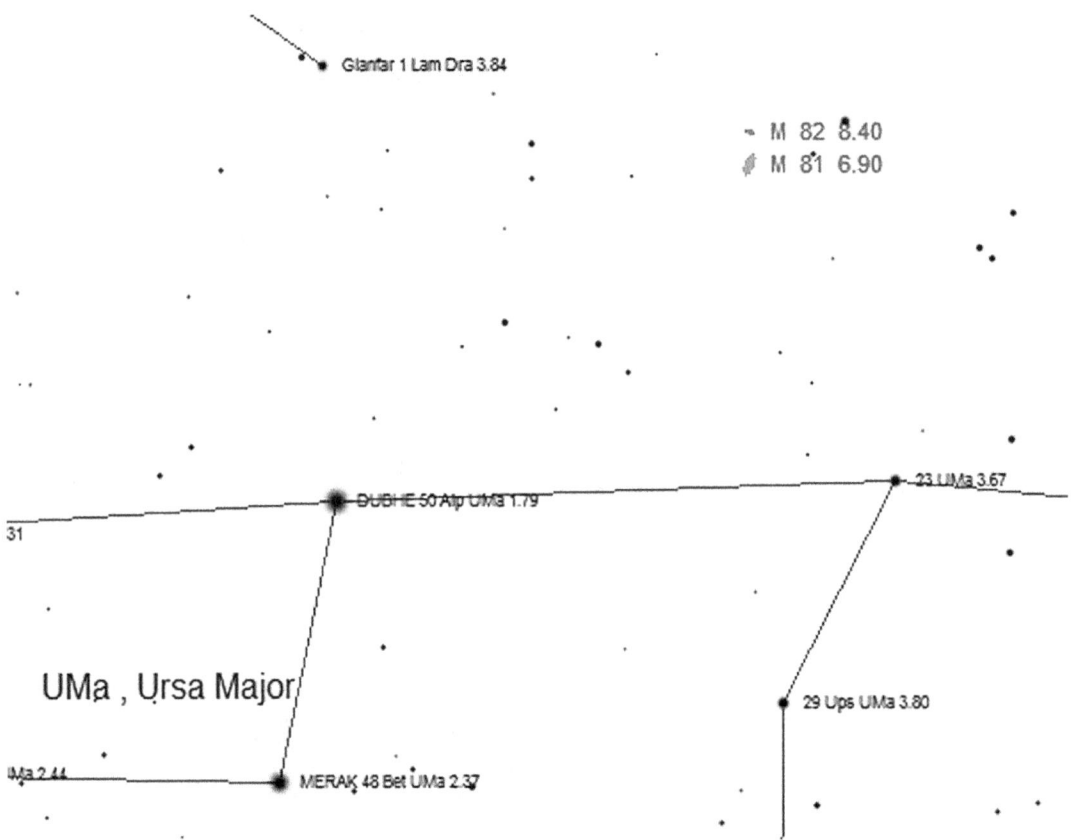

M81-82 are a pair of galaxies, bright enough to be identified with binoculars. M81 is a spiral and like all the others shows its details only with telescopes of at least 300mm (12 inches). M82, on the other hand, is an irregular galaxy from both a physical and an observational point of view: in fact, it is one of the few galaxies that show an increase in details in proportion to the instrument's diameter.

Our observations

Sketch   Observer's name:

Object and position:		Date:	Time:
Observing site:		Type of telescope:	
Diameter:	Focal:	Eyepiece/magnification:	
Sky darkness:	Seeing:	Transparency:	Moon phase:
Notes and impressions:			

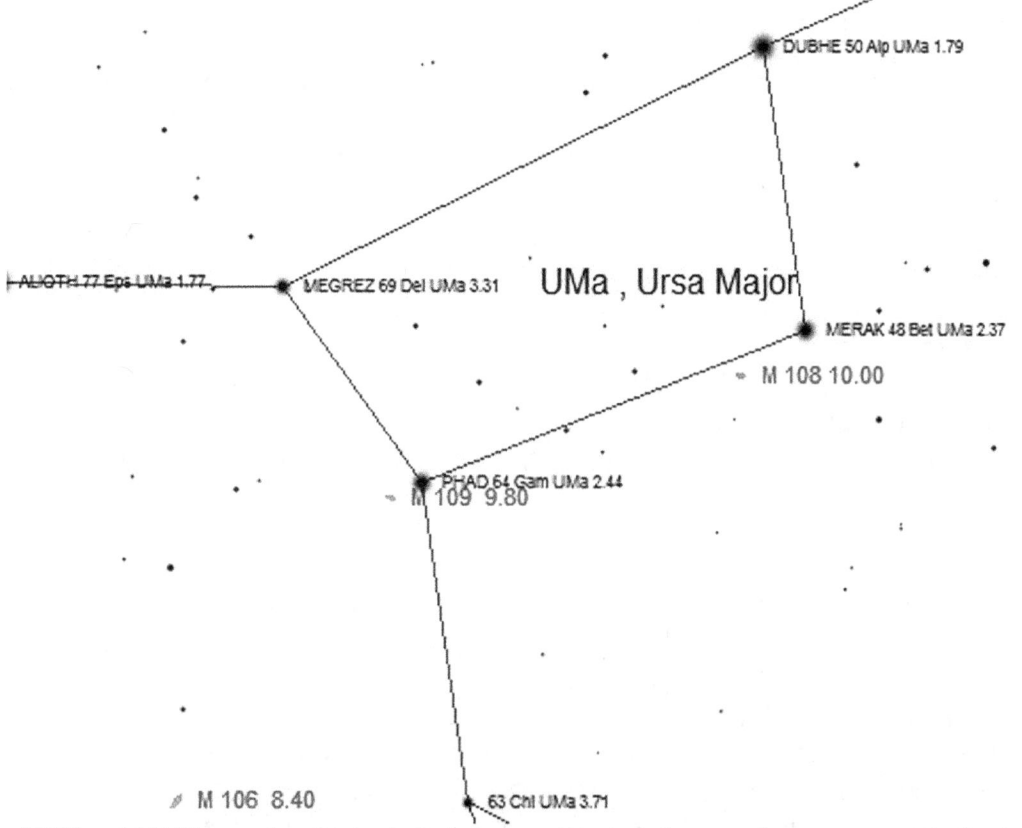

M108 and M109 are rather faint spirals, but are easy to trace because they are prospectively close to the bright stars of the Big Dipper. They're visible with instruments of at least 90-100mm (3.5-4 inches), although with difficulty. Diameters twice the size show them as evident and contrasted, although without other details which are reserved for telescopes greater than 250mm (10 inches).

Our observations

Sketch   Observer's name:

Object and position:		Date:	Time:
Observing site:		Type of telescope:	
Diameter:	Focal:	Eyepiece/magnification:	
Sky darkness:	Seeing:	Transparency:	Moon phase:
Notes and impressions:			

Some particular observations

As time goes by and as we improve step-by-step, we'll learn to observe new objects while hunting for increasingly more elusive and fascinating details. Before concluding, however, here is some last-minute advice on some observations that, while not satisfying from an esthetic point of view, could certainly excite us because of their meaning.

The colors of the stars

Stars aren't very interesting to observe because they can't be magnified and we already know this. But if we have the chance to observe a particularly beautiful double system, like Albireo, we're highly aware that every star has a different color than the others.

In a Moon-lit evening or one when we weren't able to go look for a dark spot, we take our telescope and point a few bright stars. If galaxies, clusters and nebulae are always seen in black and white, the bright stars are always colored. Other than practicing our pointing, we'll also learn another fascinating thing: the color of the stars depends, in fact, on the temperature of their surface.

The stars that seem red to us are the coldest ones of all, about 5,432°F. The orange ones, like Arthur, reach about 7,232°F. The yellow ones like the Sun are 9032°F. Then there are the perfect white ones, like Vega, which have temperatures at around 18,032°F and the last are the blue stars, like Deneb, which reach temperatures between 36,032°F and 54,032°F. Seeing the Universe in colors and also understanding how it works is beautiful, isn't it? At times even just a little is enough to have fun and be surprised.

The colors of the stars photographed with a small telescope without pursuit and a digital reflex with slowly blurring the image. Exposure time: 2 minutes.

A Colorful Universe: Star Color and Temperature
Spring/Summer

Color	Example	Surface Temperature (°C)
●	Spica (Virgo)	28,000–11,000
●	Vega (Lyra)	11,000–7,500
●	Sun	6,000–5000
●	Arcturus (Boötes)	5,000–3,600
●	Antares (Scorpius)	3,600–2,000

The colors of some bright stars in the sky. Each tonality corresponds to a certain temperature.

Alien planetary systems

There have been thousands of planets discovered outside the Solar System, which therefore orbit around other stars. It's impossible, even for many professional telescopes, to see them directly; but the idea of being able to observe a star in the image of which is enclosed even the very faint and indistinct light of some planet sends shivers up the spine.

The first exoplanet discovered (that's what they're called) was 51Pegasi b, a gassy giant that orbits around star 51 of the Pegasus Constellation. This star has a magnitude of 5.49 and is even visible to the naked eye if we have good vision. Anyway, now that we know how to use the telescope we can point it with the star hopping technique and admire it. It's another dot like many others, if we don't know what hides the first planet in history discovered outside of the Solar System.

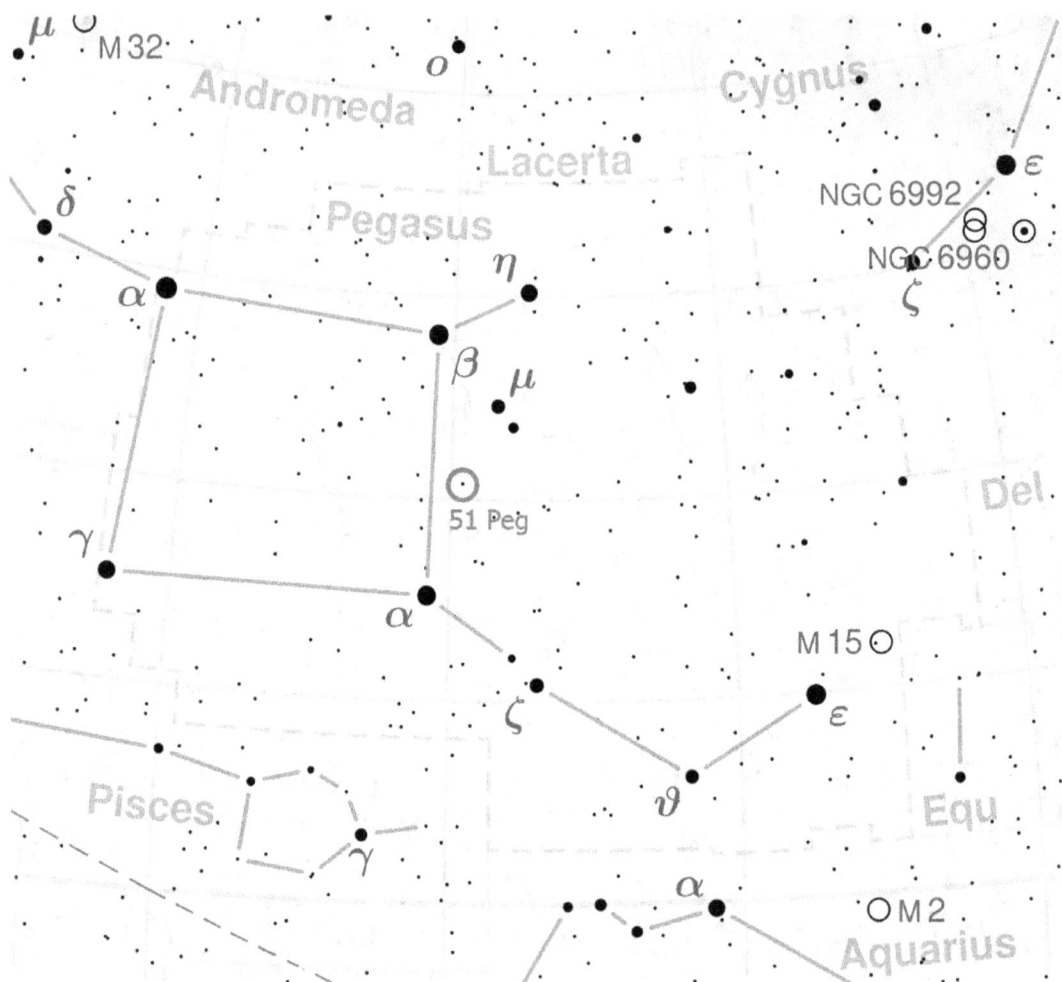

Map for identifying the 51Pegasi star, around which the first exoplanet discovered by humanity orbits.

If we want to increase the excitement level, we can choose another star around which professional astronomers have discovered a planet similar to the Earth, probably rich with water and – why not? – even with forms of life. The Gliese 667 system is composed of three tightly wrapped stars, indistinguishable with a telescope. Astronomers have discovered a good six planets around one of these stars, one of which is surprisingly similar to the Earth. The system is found in the Scorpius Constellation and has a magnitude equal to 5.89, at the limits of the naked eye's sight, but any instrument, even the telescope's finder, will show it clearly. After having carried out star-hopping, beginning with one of the stars in Scorpius' tail, we'll be able to focus on that little dot on which, maybe, other forms of life are prospering and are perhaps observing, at the very same time as we are, that sky so very different from theirs in which a very faint yellow star hides the extraordinary history of this planet and its inhabitants, fully aware of the wonders of the Universe.

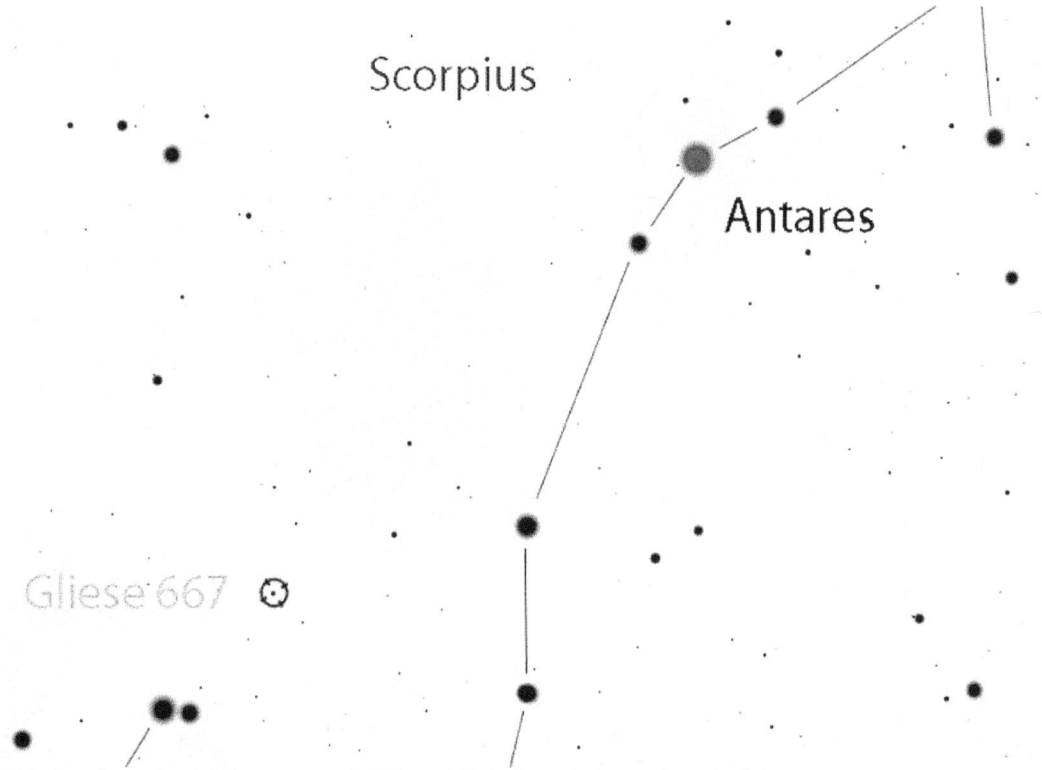

This is where the triple system of Gliese 667 is found. Around one of the stars there is a planetary system composed by at least 6 planets. One of these resembles highly resembles the Earth and could host water and forms of life. The Universe is also this: one surprise after another.

The farthest object

How far can we reach with our telescope? I don't know how many times I've asked myself this question before I found the answer. I'll bet that a lot of you have also wondered about it and I'm here to give you an answer.

The farthest objects that we can observe are, naturally, the galaxies because they are extremely large and bright. So, which is the farthest galaxy that we can reach? It's called 3C273 and it's a very particular object, named quasar: nothing more than a particularly bright nucleus of an ancient galaxy. Located in the Virgo Constellation, it's a fairly faint object, reserved for a dark sky, for someone with an eagle eye and possibly a telescope of at least 150mm (6 inches), because it shines with a 12.9 magnitude. If we can find it among a myriad of stars (a somewhat complicated operation) we'll be able to see a miniscule dot similar to all the other stars. Instead, we are observing the nucleus of a galaxy a good 3 billion light years away! Its brightness is equal to about 1000 galaxies and if it were only 30 light years away from the Earth it would shine like our Sun.

Its light, barely visible with our instruments, has completed a voyage that has lasted a good 3 billion years, crossing hundreds of galaxies placed between us. Although it may not be spectacular, it certainly deserves our respect and at least a quick thought, because we are looking at this object as it was 3 billion years ago. We have in our hands a true fossil of the Universe.

To try to find the quasar 3C273, we have to track down the Virgo Constellation and, in particular, start from the star contained in the circle at left, called η Virginis, visible even with the naked eye; then, try to hop as shown in the image below until you reach two stars close together, one of which is actually 3C273, the farthest object that we can observe with our telescope.

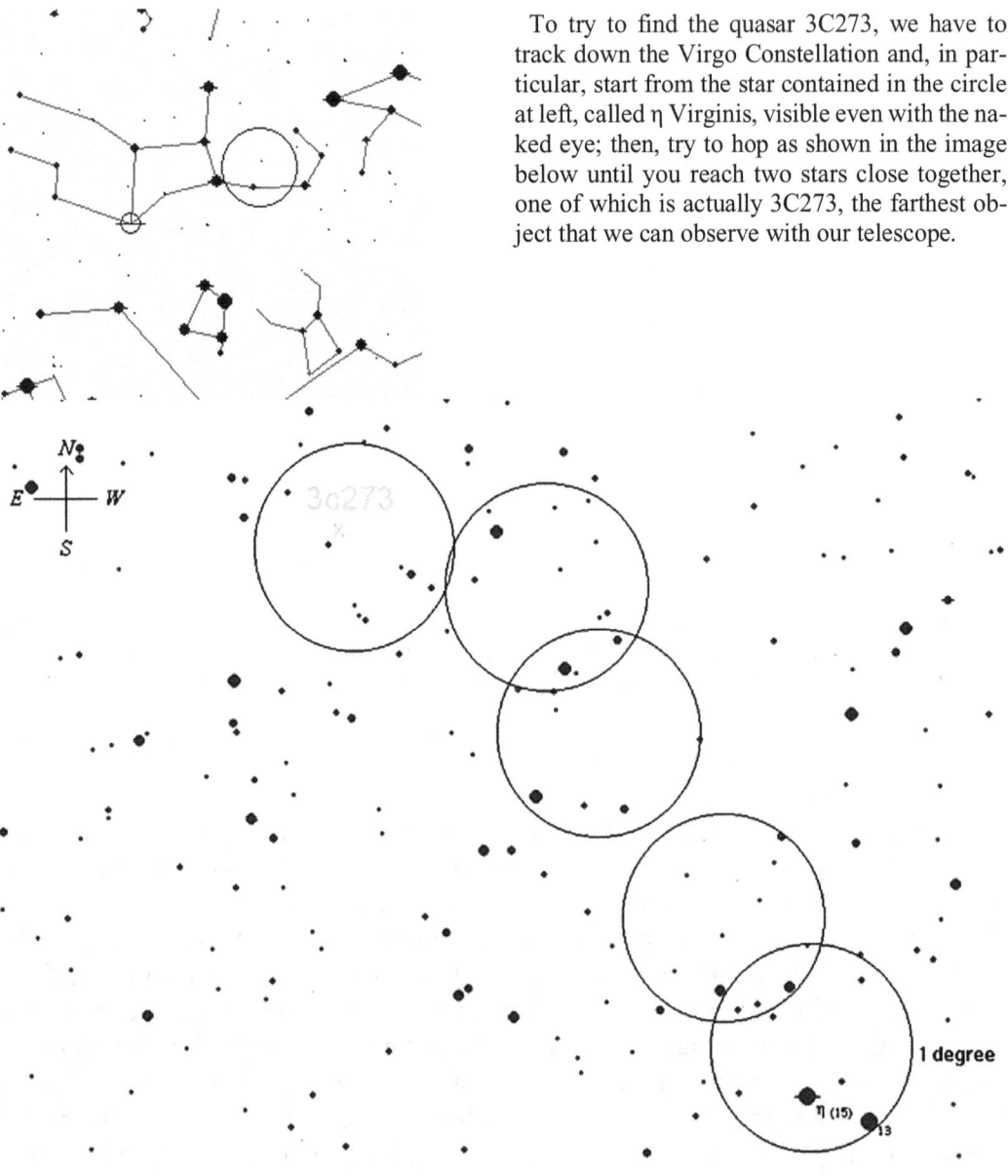

The Messier marathon

Messier's catalogue contains 110 objects spread almost everywhere in the sky, some of which are better observed depending on the seasons. One day, however, an unidentified amateur astronomer became aware that in a certain period of the year, from the second half of March to the beginning of April, it was possible for observers in the middle northern latitudes to observe all the objects in the catalogue in one entire night.

Like every out-of-the-ordinary story, the origin of the Messier marathon is wrapped in myths and legends that amateur astronomers pass down from generation to generation, but that doesn't matter much to us because we already like the challenge, even though we can face it better when we're a little more experienced.

Many astronomy clubs organize free evenings dedicated to this event which we can freely participate in or not.

The undertaking, because that is what it is, foresees observing, from sunset to sunrise, and to try to reach the mythical threshold of all 110 objects. One of the most entertaining parts is programming the course to follow, beginning with the objects closest to the Sun to then climb higher up from the horizon and ending with the last ones to be seen when dawn has already begun. It's not a race like the classic marathon, which requires a certain athletic ability, even though in some moments, especially near the Virgo cluster, you'll have to be fast enough and agile enough to quickly move the telescope and recognize those crazy balls that all seem the same. Everyone can act like they want, but the unofficial rules of this game consider the GOTO to be like doping in sports. It would be too easy: the one with the most precise and expensive mount can easily find objects in a short time and with minimal effort. And what marathon would it be like that? It would be like driving a car while all the others were running with their own two legs!

If we're enthusiastic about physical challenges bordering on the impossible, we can follow the footsteps of some extravagant American amateur astronomers (the fiercest), the greatest human GOTOs, who prepare marathons of objects that also involve the NGC catalogue, reaching the observation of almost 600 in one night, pointing them manually. I've never tried it, but this must be a much more difficult undertaking than running a real marathon!

Without reaching these numbers, we can certainly enjoy ourselves greatly with the classic Messier marathon, that is, if we choose a dark place with a free 360° horizon and no disturbance from the Moon. Searching on Internet with the term "Messier Marathon" will give us a lot of resources where we can get advice on preparing it the best way.

Keeping a record of our sky

Since we're astronomy enthusiasts and not academics, we have a freedom that professionals don't have: living a personal and unique relationship with the sky and building something extraordinary that will make us happy and improve us as people.

Keeping a diary of our observations and (why not) trying to draw what we're seeing with the eyepiece have all been a great way for me to live as an amateur astronomer.

While preparing to write this book, when I looked for and thumbed through that notebook full of photographs, writings, emotions, often errors and ingenuous questions that I had compiled between 1998 and 1999 when, at age 15, I had seriously begun doing astronomy, I was moved at seeing those ugly but, for me, unique drawings. My creations, those colored smudges were, and still are, even more beautiful than the images of the Hubble Space Telescope because they're mine, the fruit of my extraordinary experience among the stars, of a humongous dream come true and unique throughout the Universe.

I can't know if it will be the same for you. If you have read up to this point, the chances are good that it has been and so I'm speaking to you from the heart. It doesn't matter if you're 10 or 90 years old; astronomy and the Universe are truly for everyone and it's no more difficult than learning how to play with a cellphone or a videogame. It's certainly much more educational, profound and lasting than anything created by man. Because the Universe, up there, has been around for almost 14 billion years, just like all the matter from which our body is made. Let's never forget our origins, whatever else happens.

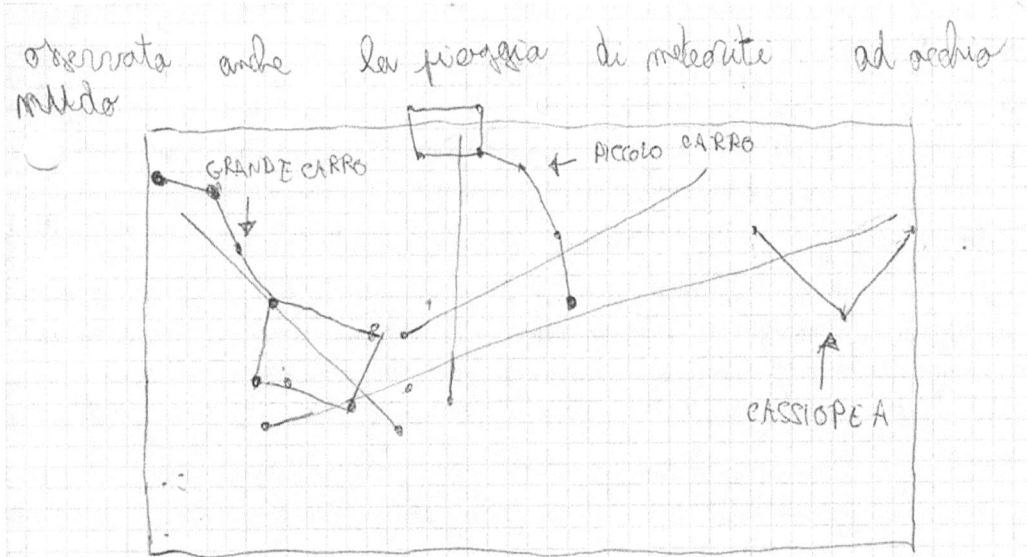

One of my observations from 08/13/1998 (sorry, I did not know English when I was 15). Keeping a diary of the sky is the best thing we can do.

When art meets astronomy

We're not talking about me, because I'm hopeless at drawing, but in observing the Universe, whoever has an artistic bent can find hundreds, no, thousands of splendid subjects to immortalize on their notepad.

The delicate filaments of some nebulae, the disposition of the hundreds of stars in the clusters and the evanescent and elegant shapes of the galaxies represent the greatest gifts that we can give to our artistic talents. In these cases, amateur astronomy melds perfectly with art, with the desire to represent a wonderful and hidden reality, according to the way we see it.

At this point, there are few enthusiasts who practice this form of art, and yet those few who resist against the cold sensor of digital cameras always have a big smile, they don't get angry trying to make the instrumentation work and they have a closer relationship with the Universe. And don't forget that they create real works of art, worthy of an art gallery.

You need at least a little talent, there's no doubt about it, but even in this case it's experience and the desire to improve that make the difference, that help us grow, improve and give us that air of happiness that never wears out when the observation is over; in fact, it lasts even during the moments of normal difficulties of daily living.

I have the privilege of knowing the best astronomy sketch artist in Italy and one of the best in the world, an enthusiast who scrutinizes the sky with his economical telescopes and has no need of a GOTO to find any object he wants in just a few seconds. He just needs a calm evening for the exhaustion from the day's labor to disappear.

Giorgio Bonacorsi gets his Dobson, takes it to the nearby hills and he's ready in 5 minutes to observe and get excited while drawing everything he can admire with the eyepiece. A little music for relaxation or the company of some other enthusiast, and the hours flow swiftly among the stars, where day-to-day problems are lost among the enormous distances in the Universe.

I've often asked him how he can do such beautiful drawings, but he answers me as though all this was normal; and in the long run, he's right.

Love for the Cosmos, not just for one's self, pure, unconditional and free of any obligations. A notebook, a pencil and an eye that tries to understand the tiniest details, the most elusive nuances, the purest beauty. It takes 40 or more minutes to do a good drawing, but only we who are reading these lines feel the time; certainly neither he nor anyone else who, losing themselves among the stars, feels the need of something to measure the time that many of us can't seem to let go of. Let's give ourselves a gift when we observe with our telescope, even if we're not as good as he is: let's leave our watch at home and be transported without another thought.

A few amazing drawings by Giorgio Bonacorsi

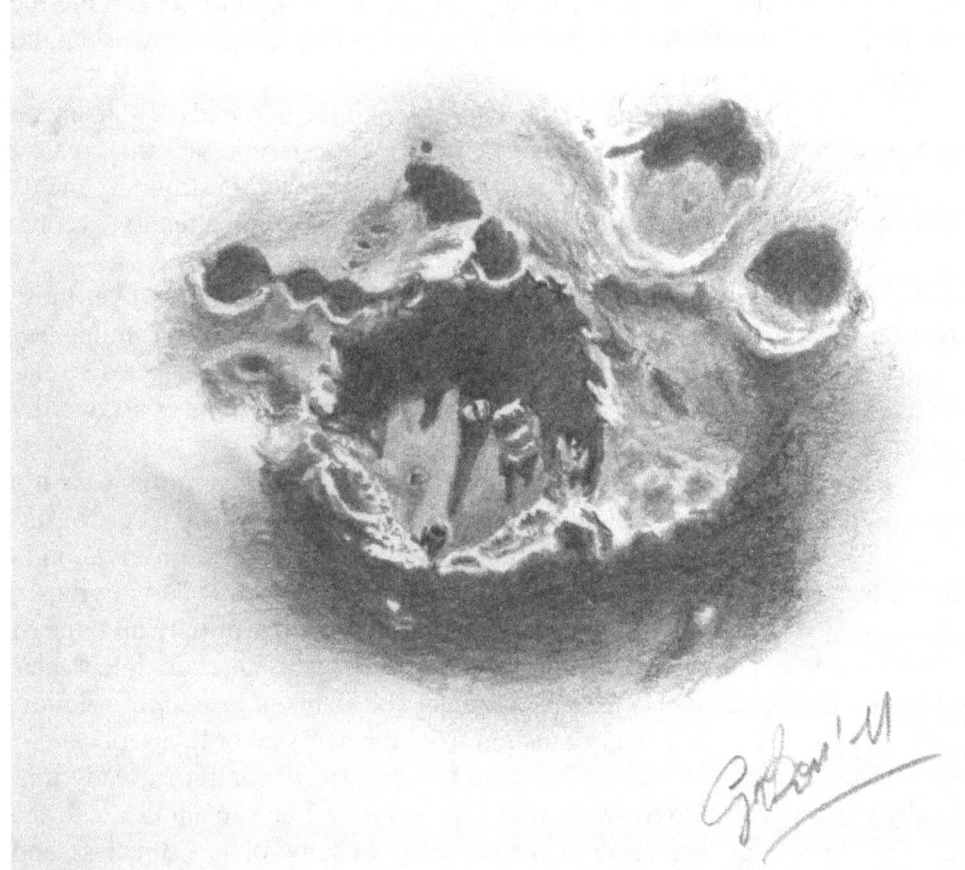

The lunar crater Walter, observed with a 130mm (5 inches) Dobson. Notice the beautiful chiaroscuros.

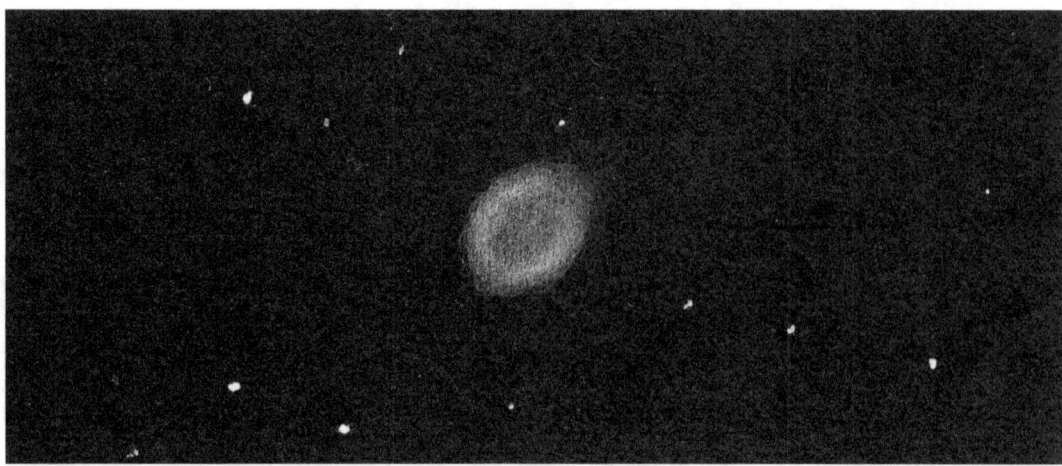

The Ring Nebula M57 in Lyra, through a 250mm (10 inches) Dobson at about 150X

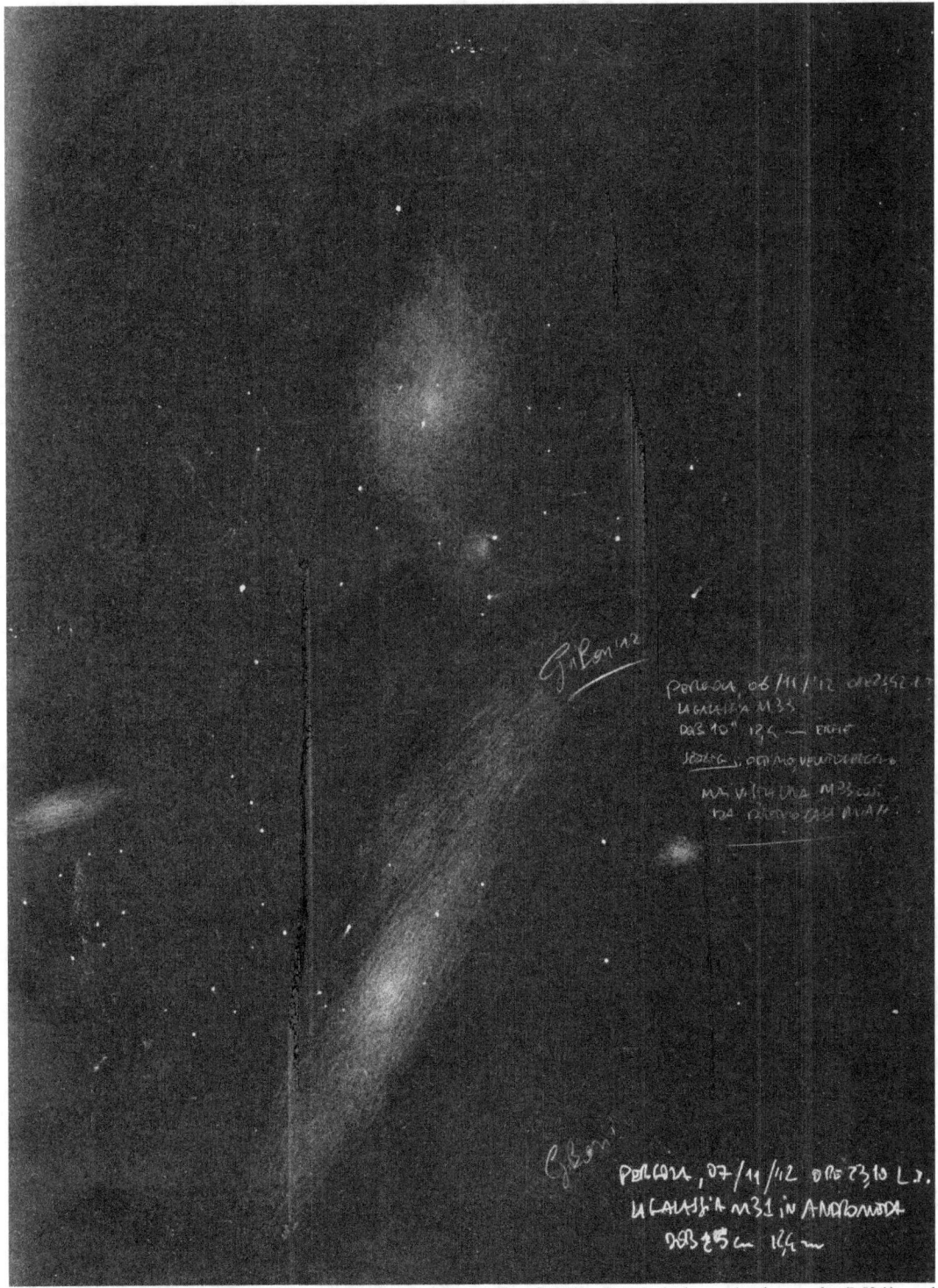

The galaxies, queens of the autumnal skies. Above: the fleeting M33; below: Andromeda (M31) and its two little galaxy satellites, observed from a dark sky and a 250mm (10 inches) Dobson at about 100 magnifications.

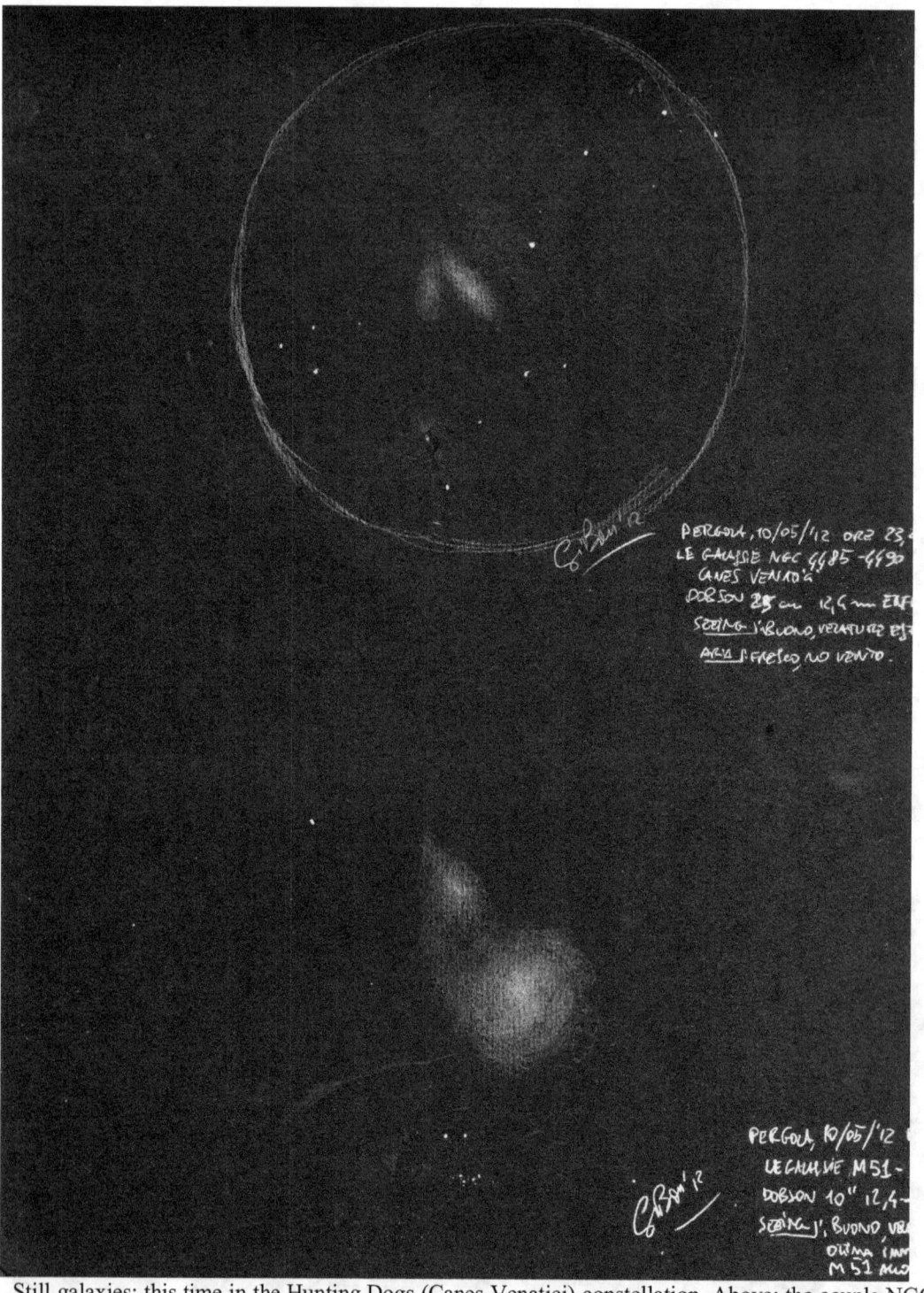

Still galaxies; this time in the Hunting Dogs (Canes Venatici) constellation. Above: the couple NGC 4485-4490. Below: the unmistakable form of M51 with its faint spiral arms. Dark sky and 250mm (10 inches) Dobson.

The Panstarrs comet, appearing in the evening sky in the second half of March 2013. Here, as it appeared with 100mm (4 inches) binoculars at 25X.

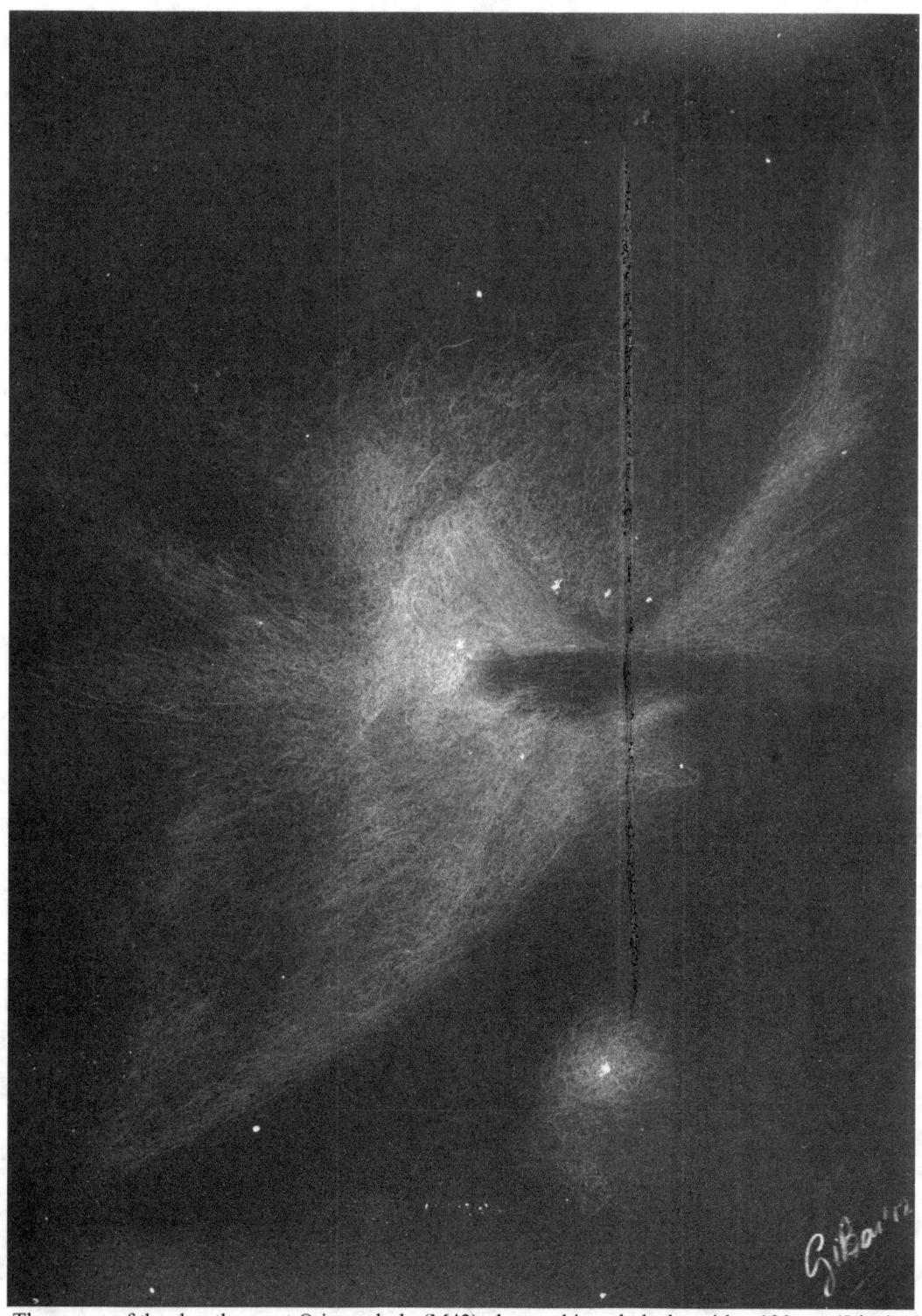

The queen of the sky: the great Orion nebula (M42) observed in a dark sky with a 130mm (5 inches) Dobson at about 100 magnifications.

Unfettered curiosity

Times have changed a great deal since I became interested in astronomy, and now the only thing blocking our enthusiasm is us.

With the possibility of using internet to download sky simulation software and all other information that we wish, all that's left is to completely unleash our curiosity and hope that a book, including this one, doesn't tell us everything we have to do. I wanted to write a small guide, but then it's up to each of us to live astronomy in the best way, observing what we like and how we like, buying one instrument rather than another. We already live in a society that expects much more from us than it gives, where the rules, at times, seem to suffocate us and block us from living. In astronomy, and any grand passion, none of this is necessary. We can read and garner information, but there are no iron rules to follow and we're free to do what we like the most. So, unleash your curiosity. Google will be your best friend during the day and the cloudy nights and remember that everyone can give you advice, but at the end of the day, no one can tell you how to live your very personal and unique relationship with the Cosmos.

Happy Universe to all of you!

Appendix

Our journey is over, but before leaving you to live your dream in peace I should give you some technical advice on maintaining your telescope. I'll spare you the details; for those, here's always the instrument's manual – certainly in a more detailed manner and many online resources.

The Collimation

If we have a Newtonian or Schmidt-Cassegrain telescope and we always see blurry images, sometimes even deformed, even at low magnifications and with very little atmospheric turbulence, then the moment has come to talk about an operation that could give new life to your instrument: collimation.

The mirrors inside the tube must, in fact, be aligned impeccably; otherwise, the images will degrade as much as it is misaligned. It's an operation that may be frightening the first few times, but if we follow the manual's instructions, there's nothing to fear.

To understand whether an instrument is collimated, we have to point an eyepiece that will give us the highest possible magnification at a bright star that's high on the horizon, in an evening with low turbulence. Let's blur the star and observe the image. If we have telescopes that use mirrors, we'll see the classical doughnut: a thick ring with a hole in the center. If the telescope is collimated the hole will be found in the perfect center of the ring. If it's not, then we must align the mirrors. In the catadioptric only the secondary mirror, set up on the connecting plate, is aligned by moving the three screws placed on its support. In the Newtons, we have to align both the primary and the secondary mirrors.

In economical refractors and Maksutovs, there's usually no possibility of aligning lenses or mirrors, but these instruments are often already perfect when they leave the factory and if they're not beat with a hammer, they don't undergo damage.

The Newtonian is the telescope most sensitive to misalignment of its mirrors and this is why we should sooner or later find the courage to learn all the phases of collimation.

Like everything else, the first time might take more than half an hour; then, with experience, it will take less than a minute to do, even before beginning an evening of observations. The instrument's manual will be our best ally.

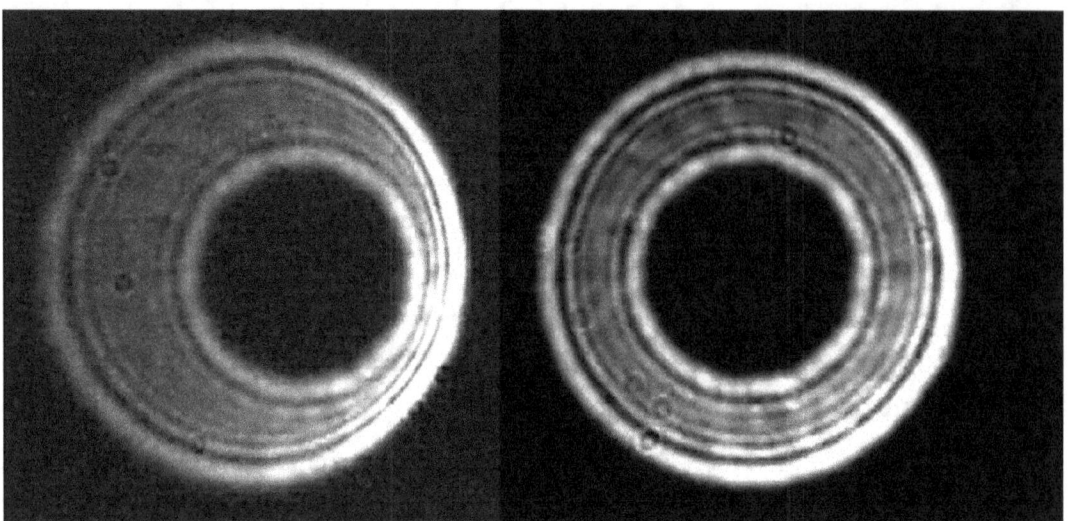

At left: defocused figure of a star that manifests evident signs of misalignment of the optics because the lack of symmetry. At right: the result of a perfectly collimated optic.

Cleaning the optics

My second telescope, a beautiful 90mm (3.5 inches) achromatic refractor and 910mm focal on an equatorial mount that seemed like it came from Mars, was a priceless jewel for me. I hugged it every day, polished its tube and admired it every night before falling asleep in my bed.

One day, I noticed that a bit of dust had accumulated on the lenses, even worse, on the internal one. I didn't how that was possible but it really bothered me so, without considering the consequences (a fairly difficult thing for a 15-year old boy to do) I decided to dismantle the objective and clean it. It was a fatal mistake, because at a certain point, after having removed the frontal lens, the one behind leaned over and disastrously fell to the ground, splintering beyond repair. I was petrified for a few minutes, and then began to think about how stupid I'd been. I had ruined my new telescope. How could I tell my parents? How could I have fixed it? I couldn't; I would have to keep it like that forever.

If this experience could have been avoided with a tiny bit more attention (even if it's difficult for a 15-year old boy), the same can't be said about the following. Yes because, not content with the damage I'd caused years earlier and which still made me feel bad, I did something else shortly after being gifted with a third telescope, costing around 3 million, 300 thousand liras (about $1800 USD): a Newton reflector with a diameter of 250mm (10 inches) and a nice tube in millboard.

That darned back mirror became dusty so embarrassingly easy, that I decided to clean it.

Dismantling it, very delicately, went well and I began to clean the mirror as I had done with the lenses: with a piece of chamois and delicate movements.

I didn't know yet that the surface of an astronomical mirror is much, much more delicate than that of a lens and when I finished, I was horrified: my very beautiful and very new mirror was full of scratches! What could I have done then? I desperately reassembled it and decided that I would keep it like that. I learned something from these two disastrous experiences that I hope will help all those who own a telescope: a little dust is normal and doesn't undermine the observation unless there is so much that you can write on it with a finger. Therefore, please, don't touch the lenses and especially not the mirrors if you don't know what you are doing; do this at least for me and for everything I've gone through!

When the time comes for cleaning the lenses and mirrors, after several years of use if we have maintained the instrument correctly (it's best to not lose the caps!); then we can think about cleaning them.

Even though we don't need it now, this is more or less what we should do. The lenses can be cleaned with an air pump that carries away a bit of dust and then delicately go over it with a piece of chamois.

Instead, the mirrors must be washed and never touched! Dismantle them from the telescope and hold them under the water faucet. A drop of neutral hand soap spread delicately with a wet finger will help take away the more resistance dirt. Then rinse under running water and then a last rinse with distilled water. The latter must be done because the mirror must dry on its own and water spots would form from faucet water.

This operation is valid for all elements that contain mirrors, including the diagonals for seeing straight mages.

Most frequently asked questions: a quick review

We've almost reached the end of this splendid journey. I don't know if it was all clear; maybe not. So, before saying good-bye I'd like to summarize many of the questions that I also asked, in the past, and I had to look for the answers, by myself, during the passage of many years. Let's look at it as a summary for clearing up any doubts. It's right that there will be others and it will be because of our desire to clear them up through our own efforts that we will make our greatest strides forward.

I'd like to buy a telescope, but I'm completely clueless on the subject; what can I buy?

Unfortunately, my advice in these cases is bitter: don't buy anything, save your money for when you're ready. In a certain sense, a telescope is an instrument for when you've "arrived"; it should be seen as the instrument that will help you put into practice the basic astronomic knowledge that you have learned. Using a telescope isn't simple;

it costs a lot of money and observations require time, experience, capability and knowledge of the sky. Without these requirements, a telescope is nothing more than a piece of furniture.

Don't use the images you can obtain as an example, because photographs and observations are two completely different things. To get a better idea, you can participate in an astronomy club near your home (there are lots of them!); usually, they're very friendly and ready to help and you can try out various telescopes to get an idea of what to expect.

I'd like a telescope with lots of magnification; how much would it cost?

The magnifications aren't the factor that determines a telescopes power, also because each instrument can reach a greater value as you desire. This is determined solely (almost) by the diameter of the objective, whether constituted by lenses or by mirrors. The diameter determines the amount of light gathered (and therefore how many faint stars I can observe) and the power to resolve (the angular distance at which I can still separate two adjacent objects). The magnification is only the means for reaching the power to resolve, determined by the diameter of the telescope.

I would like to buy a telescope; I'm knowledgeable enough but I don't know how to choose; there are so many brands!

If you're ready, then I can give you some advice: 1) you have to pay for quality; don't fall for amazing offers. An optical instrument requires highly precise workmanship and this is why it costs so much. 2) You have to know what you want to do: do you only want to observe? If yes, what? Planets? Then you need a Mak or a long refractor, at least f8, from 90-100mm (3.5-4 inches). Do you want to observe galaxies, nebulae and deep-sky objects in general? Then, get a Dobson, a telescope of which the mount is reduced to mere bones in order to favor optic power. Do you want to take photographs? Then you need a good mount, even more important than the telescope itself and for now, I advise against it. Do you want to do a little of everything? Then get a Newton f5 on an equatorial mount or a Dobson, but always avoid short achromatic refractors because they're a party of aberrations! In my opinion, you need an average instrument to begin with; otherwise you'll get tired of it quickly. A telescope of this type could be a 130mm (3.5 inches) Newton, or better yet, one of 150mm (6 inches). Regarding brands, there are some that are the same because the instruments originate from the same Chinese factory; this is why Newton Celestron, Meade, Skywatcher, Bresser, Orion, Konus, RKS and GSO are actually all the same: because the instruments all come from a couple of Chinese factories and have a better than good quality for beginners. Always be wary of unknown and super-economical brands!

What are eyepieces?

Oculars are cylindrical objects containing a group of lenses, which are used to magnify and render the image from any optical instrument, whether a telescope or binoculars, visible to the human eye. They aren't part of the telescope but are interchangeable accessories that can, therefore, be bought alone.

Which eyepieces should I buy?

Oculars are essential for visual observations; unfortunately, those furnished together with the telescope aren't always sufficient for all types of observation. Other than the focal length, the eyepiece is distinguished by the optical diagram used. The better and more complicated the optical scheme is, the better the image received will be. Don't try to save money on the eyepieces, for two very good reasons: 1) an eyepiece will last for a lifetime and can be used with any telescope and 2) if you have a telescope with an excellent optical quality but use a mediocre eyepiece, you'll have terrible images. In astronomy, the quality resulting from a complex instrument is determined by the component with the lowest quality; it's useless to have a telescope able to furnish top performances and use a toy eyepiece with plastic lenses!

The eyepieces that offer the best quality/price ratio are the Ploss, with a fairly wide apparent field and good quality optics. A good set consists of 3 eyepieces and a Barlow lens, an accessory that doubles the telescope's focal length. On average, the focals for an f10 telescope can be: 25-30mm for observing at low magnifications; 17mm for observations at an average magnification and 10mm for high magnifications.

What are filters for?

Filters help improve the view of objects, with the exclusions of Solar filters, which are essential for protecting our sight when we want to observe our Star. We can distinguish between deep-sky objects, those used for the planets and those used to improve the performance of our telescope. The first serve to reduce the light pollution at our observation site and, at the same time, to accentuate details, especially of nebulae (filters focused on the emission lines of the nebulae, the so-called nebular filters, very useful, but only for nebulae). The OIII is a typical nebular filter, focused on the main emission lines of the nebulae. The filters for planets are simple colored windows to be screwed into the eyepiece and are much less selective than the preceding ones (therefore less expensive), and are effective only with planetary observations. They select and, at times, accent the visibility of certain planetary details. A red filter, for example, gives a greater contrast to details on Mars' surface, whereas a blue one highlights the atmospheric formations like clouds and fog. A purple filter is useful for trying to observe the clouds in Venus' atmosphere. The filters for improving the telescope's performance serve to reduce the chromatic aberrations of the refractors. Practically speaking, they are filters that cut off the extreme parts of the visible spectrum and efficiently reduce the chromatic aberration, to the detriment of an alteration of the color.

Although all families of filters can be useful, it's not necessary to purchase them immediately to have optimal observations, except for a solar filter!

What difference is there between an achromatic and an apochromatic refractor, and why do they cost so much?

The laws of optics assert that a single lens isn't able to make all the wavelengths (the colors) of which the framed source is composed to flow into focus, so the blue rays will be diverted differently than the red ones, and the result is that these will be focused at a different spot than the others, helping to create an image where one color is focused and the others aren't, or rather, an image that is surrounded by colored rings. This effect is called chromatic aberration and all lenses suffer from it. To remedy this problem, telescopes formed by multiple lenses are built. The most economical solution is to foresee using a doublet called achromatic doublet. In this case, the chromatism is greatly reduced (the defect is also called this), but not completely and the images, especially if obtained through instruments with a low focal ratio, suffer a lot from this problem. Groups of 3 and even 4 lenses built from special glass are usually used to reduce these disturbances. This is how images that represent optical perfection are obtained. These instruments are called apochromatic and are among the best telescopes in the world as regarding image purity. Unfortunately, a high price is paid for this purity, from 5-6 or even 10 times higher than achromatic telescopes at the same level.

What do semi-apochromatic, ED and apochromatic mean?

Quite often there is confusion regarding this subject and sometimes retailers seem to feed this confusion. There are two methods for correcting chromatic aberration: using groups of lenses (at least 3) and an apochromatic is obtained, or else only 2 lenses made of a special substance are used. The latter solution is more economical but its results are not on a par with the apochromatic ones, which by definition are constituted by more than two lenses because they have to focus on 3 different colors, something is physically impossible with groups of only 2 lenses, regardless of how special they may be. The instruments with such a particular achromatic doublet, are called ED (Extra-low Dispersion) but aren't apochromatic but rather hybrids or, more technically, semi-apochromatic.

What are aberrations?

They are optical defects that are manifested as distortions and disturbances of the images. Some aberrations are the inevitable result of the laws of optics; others, instead, are caused by an incorrect workmanship of the lenses (or mirrors) and are generally difficult to correct once the telescope is ready. A perfect instrument can't produce aberrations on an optical axis, except for achromatic refractors with a low focal ratio.

I see everything upside-down. Is my telescope broken?

It's a very normal consequence that surprises many people. Optic laws assert exactly what you see: the image formed by the objective and magnified by the eyepiece is upside-down; that's just the way it is! This doesn't disturb us in astronomical observations because the concept of over and under is typically from the Earth, while there can be some problems in the case of observations on land. In this case, there are two accessories you can buy to correct the image: a mirror diagonal that straightens the image in an up-down image but keeps the right-left direction reversed; or else a complete straightener composed of prisms that completely correct the positioning. These accessories, above all the prism, must be used only when absolutely necessary because they degrade the image a bit. If you have a Newtonian telescope, forget this question, because you'll never be able to use these accessories on it.

I can't point objects in the sky. How is it done?

This is very common and frustrating, especially at the beginning. There are a few simple rules to follow in order to make life easier.

To begin with, align the telescope's finder. It's a very precious instrument, able to help you point bright objects with extreme ease; then bring star cards so you can easily find your bearings in the sky. An astrolabe is a very useful instrument that you can even build by yourself. If the object to be observed isn't visible through the finder because it's not very bright, don't worry: point the closest bright star and, with the star cards to help you out, draw closer to the object with small steps, guided with both the finder and the telescope, used at very low magnifications.

What is the mount for and how important is it?

A telescope's mount is something more than just a tripod like the ones used in photography. It's a support that must serve two very important purposes: 1) keep the telescope's heavy tubes stable and 2) allow you to follow the stars in their apparent path in the celestial sphere with maximum precision.

There are 3 types of mount: 1) alt-azimuth: it's a simple mount that allows the movement of two axes, one vertical and one horizontal, like any tripod in photography; 2) Dobson: a very Spartan mount that supports some types of Newton telescopes; like the alt-azimuth, it moves on two axes, but is much simpler and usually doesn't have a tripod; 3) equatorial: Other than acting as a support for the telescope, it lets you follow the stars in their path in the sky, moving on a single axis. The equatorial mount, if equipped with a motor, lets you automatically follow the stars – which won't escape from the field of view anymore – after a few seconds and represents the base for carrying out any type of astronomic photography.

My equatorial mount seems to move strangely; how do I make it point objects?
It's normal; the equatorial mount moves differently compared to all the others. Before doing any damage, carefully read your telescope's manual, which explains how to use it. In fact, before doing anything else, it requires stabilization: it must be inclined at an angle equal to the latitude of the place from which you are observing and then its axis is pointed precisely in the direction of the celestial North Pole. At this point, you're ready to use it, but you can't move it horizontally or vertically like normal tripods, because the equatorial follows the movements of the sky.

The planets seem tiny and without details with my telescope; how can I improve my view?
The planets seem tiny and without details to all fledgling observers, but truthfully, it won't always be like that. Regarding size, we're being deceived by our eye, which behaves like it does when we observe the Moon by the horizon and then high in the sky and it seems to have different dimensions: in the absence of references, the eye perceives objects smaller than what they really are. A typical example is given by the planet Jupiter which, in opposition, has a diameter of about 40". This means that a magnification of 50 times is sufficient to let us see it like the full Moon seen with the naked eye. Some will think that this is already a significant size, but actually, Jupiter at 50 magnifications will appear tiny. In planetary observations, how small the eye perceives the observed planet isn't important; what is really important is using a magnification that allows the eye to take full advantage of the telescope's resolving power and this is obtained, in an evening with low atmospheric turbulence, with a magnification of 2-2.5 times the telescope's diameter, expressed in mm.

Regarding the lack of details, there are three factors you're up against: one is certainly atmospheric turbulence. Another is the alignment of the optics (particularly if you have a reflector): unaligned optics, which means not collimated, give unfocused and fuzzy images. Last, but not least in importance, is the observer's experience: let an experienced amateur astronomer see a planet, and a person who has never looked through an eyepiece, and they will tell you two very different stories about what they have seen. Observing the fine details of a planet is a technique that is learned only with time and patience.

Which, and how many, stars can I observe with my telescope?

This question is badly put; if we're talking about the number of faint stars that you can observe the answer is several million; but if we're talking about the observation of the single stars as an activity for which the telescope is used, then the question doesn't make sense. In fact, it's true that the telescope will let you see fainter stars than the simple eye can, but it's just as true that any star seen with the telescope will always stay as a dot without being resolved.

The images of galaxies and nebulae are faint and colorless, not like they are in the photos; what can I do?

The answer to this question is: there's nothing you can do. The human eye isn't sensitive enough to perceive the actual extension and colors of deep-sky objects as they appear in photos, not even by using enormous telescopes (just think, not even if we were inside the nebula itself!)

Anyway, you can improve your view by looking for dark skies, far from the cities and keeping your eyes in the dark for at least 15-20 minutes before observing. A good trick for seeing small details consists in using the so-called averted vision, or rather, not looking directly at the object but to see it from the corner of our eye, using an area of the retina that is 10 times more sensitive.

I've heard mention about collimation and optical quality; how do I understand whether my telescope's optics are collimated and optically good?

Collimation is a procedure that is carried out on many instruments and allows the alignment of the optic axes of the lenses or mirrors in regards to the observer.

Aligning the optics, if the telescope allows for it, is an essential (but easy and risk-free) operation for being able to have better images, especially when high magnifications are being used.

Instead, an instrument's optical quality has nothing to do with its collimation. This represents the quality with which the optical elements have been crafted. An instrument that is of a low optical quality produces images that are poor in details and always blurry, regardless of its collimation, and can't be corrected in any manner. To verify the quality of an optic, carry out the star test and analyze the images of both an unfocused and a focused star, using high magnifications. But if you have bought a telescope from one of the brands I've advised in the text, from a specialized retail seller, this problem doesn't exist.

Is a lunar filter necessary? Where can I buy one?

A lunar filter is a simple dark green filter to be screwed onto the eyepiece and helps to screen the lunar light when our satellite is close to its full phase. Contrary to the observation of the Sun, observation of the Moon isn't damaging for our sight but can be disturbing if conducted in the full Moon phase, using low magnifications. The lunar panorama is illuminated more or less like a normal daytime landscape here on the Earth, and for an eye used to the dark, with dilated pupils, this produces an effect similar to turning on a strong light after having been in the dark for a few minutes: annoying but not harmful. A lunar filter isn't a priority but a convenience and any colored filter for the planets can be used for this purpose.

My optics fog up when I observe; what can I do?

It's a common problem, especially for telescopes with corrector plates like the Schmidt-Cassegrain and, to a lesser degree, for refractors, while it isn't for Newton reflectors (but not always; sometimes the secondary mirror fogs up when there is too much humidity). The first thing to do is prevent: in evenings with high humidity, you need to build a long hood from black cardboard with which to envelop the anterior part of the telescope. If this isn't sufficient or if it's not possible to apply it, and the optics fog up, the only way to dry it is to obtain a hairdryer, with a cloth in front of the vent to block the dust, and to blow warm air over the fogged-up surface. Never try to dry the optics with rags for cleaning windows or with products for glass; otherwise, you risk ruining them.

Venus, Mercury and Mars observation sheet

YEAR: _____ MONTH: _____ DAY: _____

OBSERVER: _____

LOCATION: _____

START OBS.(UT): _____ END OBS.(UT): _____

TELESCOPE: _____ DIAMETER: _____

F/ ___ MAGNIFICATION: _____ FILTERS: _____

SEEING: _____ TRANSPARENCY: _____ COMFORT: _____

(UT): _____

PHASE: _____
NOTES:

Jupiter observations sheet

YEAR: _____ MONTH: _____ DAY: _____

OBSERVER: _____

LOCATION: _____

START OBS.(UT): _____ END OBS.(UT): _____

TELESCOPE: _____ DIAMETER: _____

F/___ MAGNIFICATION: _____ FILTERS: _____

SEEING: _____ TRANSPARENCY: _____ COMFORT: _____

(UT): _____

NOTES:

Saturn observation sheet

YEAR: _____ MONTH: _____ DAY: _____

OBSERVER: _____

LOCATION: _____

START OBS.(UT): _____ END OBS.(UT): _____

TELESCOPE: _____ DIAMETER: _____

F/ ___ MAGNIFICATION: _____ FILTERS: _____

SEEING: _____ TRANSPARENCY: _____ COMFORT: _____

(UT): _____

NOTES:

Saturn observation sheet

YEAR: _____ MONTH: _____ DAY: _____

OBSERVER: _____

LOCATION: _____

START OBS.(UT): _____ END OBS.(UT): _____

TELESCOPE: _____ DIAMETER: _____

F/ ___ MAGNIFICATION: _____ FILTERS: _____

SEEING: _____ TRANSPARENCY: _____ COMFORT: _____

(UT): _____

NOTES:

Observer's log

Name:

Entry 1

Object and position: Date: Time:
Observing site: Type of telescope:
Diameter: Focal: Eyepiece/magnification:
Sky darkness: Seeing: Transparency: Moon phase:
Notes and impressions:

Entry 2

Object and position: Date: Time:
Observing site: Type of telescope:
Diameter: Focal: Eyepiece/magnification:
Sky darkness: Seeing: Transparency: Moon phase:
Notes and impressions:

Entry 3

Object and position: Date: Time:
Observing site: Type of telescope:
Diameter: Focal: Eyepiece/magnification:
Sky darkness: Seeing: Transparency: Moon phase:
Notes and impressions:

Entry 4

Object and position: Date: Time:
Observing site: Type of telescope:
Diameter: Focal: Eyepiece/magnification:
Sky darkness: Seeing: Transparency: Moon phase:
Notes and impressions:

Biography

Daniele Gasparri was born in 1983 in a small town in the center of Italy, where the sky was still dark enough to see the Universe. He first observed the Moon when he was 10 and it was love at first sight.

He started taking pictures of the sky in the late 90s, before attending the university and graduating in Astrophysics and Cosmology at University of Bologna.

In 2018 he realized his dream: moving to Atacama desert, where the sky is so dark that the Milky Way can cast shadows on the ground. He is actually doing a PhD in astronomy and planetary science at Universidad de Atacama.

Daniele is also a professional science writer and communicator. He published 35 books about astronomy and many outreach articles in astronomy magazines all over the world. This is the first book translated in English!

From the pristine sky of the Atacama desert, Daniele is continuously looking for the beauty of the Universe. The marvelous colors, the smooth contrasts, the delicate shapes, the unimaginable distances… they just need to be captured by a camera and shared with the world.

To know more about Daniele, see his website: www.astroatacama.com